# SHAKESPEARE

# Shakespeare

## *The Reformation in the Form of Plays*

PETER RAJ

Peter Raj

First Printing, 2024

# Contents

# Chapter 1

# William Shakespeare

Was the Bard of Avon an existentialist? Selective Retrospective Existentialism is applied to William Shakespeare. We find recorded in history the year of birth (1564) and death (1616). It is in Hamlet we hear the cry "Adieu, adieu, remember me." It was on BBC television he was being remembered. I sat down to watch the story of the man who wrote Romeo and Juliet. The only other thing I knew about the greatest writer of all time was that he was hard work. I needed Shakespeare-made-easy. "Shakespeare: Rise of a Genius" seemed to promise that. I was willing to watch all three episodes for the pleasure of knowing why this man is so revered by theatre goers. The programme was in celebration of 400 years of his plays. They were published in a book called "The First Folio". The rest as they say is history.

"Oh my God" I said aloud in astonishment. The word "existentialist" jumped out of the screen when Hamlet appeared. It was Shakespeare's famous words resonating with me, "To be, or not to be?" It is a question of existence and it prompted me to go to the play. Reader look, not to his picture, but his plays! David Tennant was in the title role and after 25 minutes I was favoured with another existentialist line. "This above all - to thine own self be true." This was elaborated. "Thou canst not then be false to any man". It is quite simply the first principle of existentialism, and it is found in Hamlet, a play written more than 400 years ago. "To thine own self be true" was quite a challenge back then. England was drowning in creeds, but existentialism was not a creed. Genetics and the selfish gene like evolution and Big Bang did not figure in the truth of the day. Polonius ought to have said to thine own self be true as best as is possible.

Another burst of existentialism came in King Lear. It ends with, "Speak what we feel, not what we ought to say". In other words, be real. There is existentialism in Othello to rival Hamlet. From the existentialist point of view the character of Othello is pure evil. He is deception personified. Honest Iago is at least honest. He declares "I am not what I am" and that is the basis of his honesty. If we are not seeing the real Iago, then we might have to go beyond the surface to find him. We might even have to go outside the play. Every Elizabethan watching the play for the first time would have been shocked by the presence of a Moor on stage. Hate and suspicion was not unique to Iago. Meanwhile, Desdemona invests her emotions unwisely. She is the existentialist because she is true to herself, speaks what she feels, and she begged to exist for another day. The Moor could not be reasoned with. He had a sense of self-worth which was nothing short of delusion.

Let Desdemona's tragic demise be a lesson to us all. Iago was right to hate the Moor. Shakespeare had to be mindful of religious sensitivities. Truth is not possible when people are taught supernatural dualism. Such people believe they are part biological and part spirit. They are part eternal, and that part is the problem. Truth becomes blasphemy. Dualism defines them and their universe. Christianity is one of many operating systems for that dual universe. But time came to review the framework. The first article was whether English king or

Roman pope was best placed to serve Christian England. Protestants and Catholics were suspicious of each other. The Reformation dictated Shakespeare's editorial policy. The situation demanded intellectual discipline. Shakespeare knew not to upset extremists. Othello is as much about what is not said.

It was episode 2 of "Shakespeare: Rise of a Genius" which introduced Hamlet and I became an ardent fan of the playwright. Conscience is the theme dominating that revenge story. It was four hundred years later, and I was celebrating Hamlet's calculated revenge on death. It's like he focused on it knowing there would be a reaction on my part. We know the author by his plays, and he knows us by his plays. But there is another source of information about him. Our knowledge of Mary Arden (c 1536 - 1608) is our knowledge of William Shakespeare and the fact that she made him and got him into grammar school says a lot about her determination to succeed. She was readying him for university. She shared the secret of success with her son, and he learned all about commitment and hard work. Otherwise, the story would have been the crushing of genius. It must have been humiliating to lose his place in school. By 13 the door to the grammar school was firmly shut behind him.

Christianity was not working for young Shakespeare. It was not shielding him from life's slings and arrows. His even younger sister Anne was buried on April 4th, 1579. It felt like the war with the devil had been lost. Death struck Hamlet and he asked Horatio to remember him. Living memory or Selective Telescopic Existentialism was where Shakespeare could be found after his death. He was living memory in his daughters Susanna (1583 - 1649) and Judith (1585 - 1662). He took up a considerable amount of memory in his friends John Heminges (1566 - 1630) and Henry Condell (1576 -1627) who gave us "The First Folio". In the preface they listed names from the Globe Theatre Company with memory. King James was a source of living memory till 1625. Hamlet's plea for remembrance was to the theatre family. The theatre audience was his extended family. Shakespeare had one family in Stratford and another in London.

Shakespeare wrote plays which were popular. The stage was a place of transformation and escapism. People got quality entertainment for their money. They got a visual spectacular. They picked up incomparable soundbites like "Free of speech". He raised the bar. He gave them history plays and comedies which were ever more immersive. They could talk issues without taking sides because they were talking about a play. He was generating a new consciousness. Most of the population was illiterate and the only history they knew was that of the Catholic God and the Roman Catholic Church. The Spanish Armada and the gunpowder plot left no one in doubt as to the explosive times they were living in. Audiences fell in love with history. Shakespeare learned to circumvent philosophy and only an existentialist could have done that. The name of the game was increased consciousness. It was to be achieved through history and not science.

Theatre was carrying out a social experiment. It was engaged in a subtle form of social engineering. Shakespeare chose words and scenes to make people more aware of the world they were living in. There was more to life than philosophy and church imagery. Bread and wine which turned into the literal body and blood of Christ was a sacrament and sacraments made for a philosophical civilisation. No one in Elizabethan England was challenging that directly. Reformation and Protestantism was about the supremacy of the pope which Henry VIII contested in 1534. The Act of Supremacy declared the King was "the only Supreme Head on Earth of the Church of England". The Church of England was formed out of the Roman Catholic Church which itself was established by Emperor Constantine in AD 325.

The Nicene Creed of 325 and the authority of the Catholic Church went mostly unchallenged till 31 October 1517 when Martin Luther published his "Disputation on the Power and Efficacy of Indulgences". The reformer asked, "Why does the pope, whose wealth today is greater than the wealth of the richest Crassus, build the basilica of St. Peter with the money of poor believers rather than with his own money?" The church was a successful business. The pope was an oligarch. Business expanded to include licences for pleasure. The church could forgive sins for a price. Reformers believed corruption could be weeded out and the faith kaleidoscope reset to prevent a repeat. Reformers looked to the church for direction. Henry's reformation had a lot to do with divorce and remarriage. It was another strain of reformation. By the time of William Shakespeare arriving in London the Reformation had not resolved itself.

Society was still changing and coming to grips with the phenomenon. The plays did not ignore the Reformation and in their own way added to the debate. They could not help but increase consciousness of one thing or another. It was all leading to a better-informed public. If the plays made a statement, it was that there is more to life than philosophy. Meanwhile, Catholicism thrived in the realms of decreased consciousness. Catholic priests could assume moral responsibility for everyone else and for a long time that worked for everybody. Shakespeare created thought provoking plays which encouraged people to imagine moral responsibility and safely pass judgement on fictional characters. Iago is the most unpleasant character in all of fiction. Iago is the most controversial and thought provoking of Shakespeare's characters. It took the greatest writer of all time till 1593 to get into his stride. He published a poem. One year before that he was the subject of criticism.

The earliest copy of Hamlet is dated 1603. Therein London received a huge dose of existentialism. Theatre was first to express existentialism. Hamlet the existentialist play mirrored existence as it was being experienced by the author. He lost his son who was buried on 11 August 1596 and his father who died on 7 September 1601. The one thing about existence we can be sure about is that it is always interrupted by death. Shakespeare loved his hometown. Born and died in Stratford-Upon-Avon. His family was always there and drawing him back. The emotional connection with the past remained a big part of him throughout his life. He was a romantic. He was an existentialist. His exclusion from school was a stigma. Better opportunities awaited him in London.

Mary Arden taught William all about hard work and he was not afraid of it, but hard work was killing his talent. Once he knew something inside him was dying it was all about rescuing the talent and making something of it. He arrived in London in 1587 looking for distraction, looking for the compelling. He needed to rekindle his love of writing. He published a poem, but theatre was his true calling. He wanted to take his writing to another level. He did and he made a name for himself. "Adieu, adieu, remember me" in terms of Selective Telescopic Existentialism was Hamlet's request for Horatio to remember him. When living memory expired it became Selective Retrospective Existentialism to recall him. Dame Judi Dench cannot remember the man for not having met him, but she can draw on the past and honour his name. It is essential to have some history and ideal to have it all. Hamlet was haunted by his past. Revenge and conscience are employed equally existentially in that great work.

The 16$^{th}$ Century began with everyone in England a Catholic. The Reformation in Europe was bound to have some effect. Henry VIII instigated his own Reformation. The population was forced to choose between the leadership of the Catholic

Pope and the new administration of the Church of England. They found they could live with the change. Shakespeare was a Catholic forced to become a Protestant or at least pretended to be one. Either way, religious issues affected everyone. The Reformation increased consciousness of philosophy or what we today call theology. In Hamlet we find belief in the supernatural, the play includes afterlife and Purgatory. But Hamlet's last words, *"The rest is silence"*, hints afterlife is not a natural hope. The longer the Reformation went on the more complex it got.

The University Wits of the Rose theatre included atheists. The Christians of Stratford would not have been impressed with the company Shakespeare was keeping. As far as they were concerned, he was being pulled in the wrong direction. They did not want to know what was going on inside theatre. They suspected all kinds of scandalous behaviour. Fact is, when Hamlet becomes conscious of conscience (which makes *cowards of us all)* it is a shift to innate dualism. Conscience is self. Conscience begins to replace faith in the world of Shakespeare. All started by Luther's distinguishing of Christ from the church bearing his name. Stratford-Upon-Avon was a reminder of his Christian roots. His mother and father and wife and children were never too far away from him although far enough. His sister Anne lay in a Stratford grave. When Shakespeare left for London, it was with one life already lived out.

It was not till he lost his son that his whole world caved in. Out of this personal tragedy Hamlet the tragedy was born. His art form enabled him to translate the tragic event into a play. Shakespeare lost the meaning to his life. Tragedy and not science broke the illusion of an all-caring creator God. He was not first to ask, "Why hast thou forsaken me". He forsook the metaphysical God which wasn't there and philosophy which no longer made any sense. Big Bang would be the next theory of everything, but he could not refute the supernatural because the science did not exist. In fact, in his day metaphysics was considered theoretical science. As for Christ, death or as near as, completed his integration into the world of man. The world's greatest researcher would have known that although Henry achieved independence from Rome the church had awarded him a title all his own, "Defender of the Faith". He was assured of a place in church history.

The real reformer in the royal household was Catherine Parr who was wife number 6. She passed the flame of Protestantism to her stepdaughter and ward, the future Queen Elizabeth. Shakespeare the existentialist flourished during her reign. The playhouse offered rivalry to church. There are 2 ways of destroying philosophy. One is science and the other is history. Shakespeare used history to drip-feed existentialism into the ears of his London audience. Hamlet received instruction from a ghost which to atheists was a literary device but to the faithful a part of overarching reality. It was not a ghost which offered Horatio method. *In this harsh world draw thy breath in pain to tell my story. The rest is silence.* Hamlet needed for his friend to tell his story to keep his memory alive. Beyond Horatio's living memory his turn at existence was over. Heminges and Condell gave us the First Folio. By publishing the plays, they made sure we remember ideas which came out of their age but are for all time.

By not publishing stories about the man and keeping personal memories to themselves his friends left nothing to distract from his literary achievements. It is like the author's own story is not to go beyond them, not to go beyond living memory. The BBC had to put together a background story which they did using the plays. Hamlet was inspired by the death of his son Hamnet. King Lear revealed a lot about his relationship with his daughters. They seem to be reasonable assumptions. His family naturally meant a great deal to him. It is no coincidence that the First Folio was published the same year Anne

Hathaway died. She was being protected not so much from scandal as the celebrity status her husband had achieved. She had married a crow who turned out to be a genius. Shakespeare the absent husband and father was not looking for vindication nor wishing to change her perception of him. If we are to remember him for his ideas, then we are to remember him for existentialism.

Those of the faith family have religious heroes to remember. Remembrance honours those we care to remember. Shakespeare's theatre friends thought about him constantly between his death and 1623. They were free thinkers and religious rebels who had managed to escape attention and persecution. Theatre was their thing. The University Wits attracted like-minded graduates. They could pay handsomely for the best minds. We know of atheists, homosexuals, and squanderers. We do not know of earls and lords being in the same theatre company. As part of the nobility, theatre was considered beneath them, and they naturally wished to avoid association. Those with an existentialist bent end up resisting the faith family. They have their own family. The hate of faith has never been more strikingly expressed than in Othello which is about a man with a dark faith. Othello did not appreciate existentialists. He was a man committed to re-writing history.

Existentialism was not a creed till after the Second World War. In the 1590s it was a new stream feeding into the river of Protestantism. The principles of the newly born ideology are there in the works of William Shakespeare who left Stratford-Upon-Avon a Catholic. By 1603 he had developed a new system of ethics. His plays are about individuals who are individualistic. His plays are about history and forming a strong connection with the past. Luther pursued a stronger connection with the historical Christ. Philosophy only wants that from the past that which justifies faith. Othello The Moor of Venice made some members of the audience shudder. The sight of Othello on stage was unnerving because philosophy was all that mattered to him. He was a man who showed great contempt for history. The existential did not matter to him. Philosophy, especially his own philosophical civilisation, was good reason for him to wipe out history which challenged his beliefs.

Othello reminded the audience of ethnic cleansing, social engineering, and cultural eclipse. Rome in her time civilised barbarians. The Moor did the same. Having run out of barbarians the two empires were left to civilise each other. Othello was the product of Moor Imperialism. His philosophy reached different conclusions. In his own kingdom he was free to declare Christ had not died on the cross. Similarly, he could reject a Jesus who was as big as the solar system. In the Christian empire he was allowed to pass himself off a Christian on basis that he acknowledged a historical Christ who was a prophet. By the time the Moor appeared in Othello it was too late for the two civilisations to compromise or reach a middle ground in philosophy. Blasphemy laws stopped the Moor preaching in Christendom and in the Moor's own empire blasphemy laws stopped Christians preaching. Atheists and existentialists dared not assume freedom of speech in either empire.

For the sake of peace and to end perpetual war, an unwinnable war, the two enemies had to learn to respect each other's beliefs. The Moor and the Christian when in each other's company went quiet on certain issues. They developed protocols to avoid argument and antagonism. In time there was a Moor in Venice. For the sake of the play Othello the able general is found assimilated by Venetian society. Catholics and Moors learned to hide aspects of their philosophy from each other. They were both used to hiding history. The 20th Century saw the Christian church hiding science and hiding from science. Philosophy became all about hiding the truth and hiding from the truth. Theology ran out of cover in 2010 when Science pronounced philosophy dead. Anyone who revises history for the sake of religious belief is the enemy of existentialism. Othello served Venice and served her well. England might have been entertaining the idea of an alliance with the Moor.

England was thinking of getting into bed with a Moor. Shakespeare was able to warn England of dire consequences with a play about sweet Desdemona. An unforgettable and haunting play about naivety, terror, and brutality. Quite simply, philosophers and theologians whether Christian or Moor cannot be trusted. Audiences in evangelical meetings must expect to be deceived and philosophers for their part will maintain the deception for as long as possible. It is their profession and a career. It is an oversell. Church audiences need to defy church guidance and find history, and science for themselves. The fact is that Christ instituted church and it ought to be a good thing, but it is not. The apostle Paul took it over. God the non-Jew was overruled by a patriotic Jew and after him a philosopher Moor. Both deprived history of God.

Christians need to have hate and revenge back on the moral compass because only then will they see how badly philosophy has distorted their humanity. They are afraid to hate because it is not a fruit of the spirit. They will not admit hating yet they threaten me nuclear annihilation. It must be out of love. Apparently, the Russian Orthodox Church allying itself with the Russian Federation is to consolidate Russian forces in the war against the devil and his hordes of demons. Demons have infiltrated Eastern Europe. The fascist enemy is led by warped philosophy but then it is all warped. Propaganda is nothing to do with reality just as faith is nothing to do with reality. There needs to be an existential concept of church for future generations. Church inclined people of today need to come together as existentialists. The poor Russians who have had to flee faith and propaganda will not want to return to an outdated system. The architects of destruction must be kept away from the future. Shakespeare attacked all philosophy with the line, "I hate the Moor".

The Bard introduced existentialist principles. His source was reality which we all experience. Human beings have always been existential creatures because that is what we are. The product of evolution, animals. The faithful claim the just live by faith which is evidence of things not seen but as we now know things not seen because they do not exist. They invented socio-spiritual truths and institutionalised them using philosophical methods and mechanisms. Those who say we are half spirit, and half biology are deceiving themselves and everyone else. They see a spirit being where there is none. They devised a working system which makes the religious superstition credible. Shakespeare was a force of nature, and his power was verse. People enjoyed the acting out of his plays as much as we do in the 21st Century. Difference being, they appreciated Shakespeare as the source of existentialism.

# Chapter 2

# Elizabeth I

Thank God we are no longer brainwashed from an early age by the wicked agents of the Abrahamic faiths. We have broken free of philosophy's culture grab. We just need to be honest about spiritualisation. How many believe in an omnipresent and all-pervasive Jesus Christ? St. Paul was a revisionist. He re-wrote history so that it would conform to the Word of God, a product of philosophy. Othello the Moor did Protestants a favour by showing how easy it is to re-write history. He made it say Jesus Christ did not die on the cross. Not many people realise that history can be used for re-writing history. Romans had their empire and Roman law. Anyone guilty of sedition was crucified. If we can work out Christ's sedition, we will arrive at the historical Christ. If the New Testament writers had put the same emphasis on history as Shakespeare, they would not have needed another source of revelation. "Christ after the flesh" was basis for cancelling the things said by Christ. St. Paul could not use quotes from Christ in his letters.

The Reformation in Europe which was at first about "indulgences" became about the authority of the pope. It was about the papal interpretation of the Word of God. There was always more than enough Protestantism going round to make everyone acutely aware of the friction between Christs of theology and history. The importance of history was appreciated by Shakespeare. Catherine Parr had ideas which came between Henry and his Defender of Faith title. Her death warrant was signed. What were those ideas which almost got her killed? We know we cannot trust theologians because they hide truth which contradicts philosophy and undermines faith. They know how to coerce when they need to. Some of us know the only way to fight faith is to fight dirty. We can go much further than Martin Luther by demanding faith be rooted in Christ's history and not the Word of God.

Christ of history and Christ of the natural world and Christ of Catholic philosophy and Christ of St. Paul's theology left many a Protestant confused. Indulgences was more than a money-making scheme; it was evidence that the oligarchs running the Holy Roman Empire had lost their grip on spirituality. Church was cynical and exploitative. Bishops exceeded authority and were guilty of usurpation because Christ was never like that. It was time to return to the original Supreme Head of the Church of Rome and for that history was key. History which was record of real-time and not the history of faith. Catherine Parr in 1546 submitted Protestant ideas which shocked the newly formed Church of England of 1534. Her idea of history did not include evolution of the species or Big Bang. It did include God in the person of Jesus Christ. It was history without philosophy, meaning much more existential than previously thought possible.

The Church of England was loathe to move away from the eternal society run by her philosophers. We saw much the same thing in 2006 when Richard Dawkins discovered there was another way of reading the Bible. The Church of England was unmoved by the God of War. If we follow Catherine Parr, we end up in the world of the First Folio. If we follow Dawkins, we end up in the world of the God delusion. Catherine Parr pointed the way and Elizabeth Tudor was her most faithful student. In which case spear shaking preceded Shakespeare making him a spear in the hand of a queen. We do not know anything of Catherine-Athena's will other than that which is captured in the plays inspired by her. We know it involved nature, science, history, conscience, and innate dualism as opposed to letting faith and philosophy overwhelm her notion of reality.

Church is useless if humanity is unable to transfer moral responsibility to priests. Christ of the church is useless if it is not possible to transfer moral responsibility to him. If the point of moral beings is to be moral, then church is a big obstacle. A person needs to make choices and not have them made on his or her behalf. The amount of people who believe in the existence of demons is boosted by churches. Belief in prophecy and predestination are similarly boosted. A way of thinking is all that it is. Pilgrim must retain control of his or her thoughts and actions and be able to monitor progress of character. Christ was righteous when threatening war against Rome. Circumstances changed and so did righteousness. Christ was righteous when withdrawing from politics. His kingdom is not of this world. His kingdom is off world. Righteousness is fluid. The ten commandments are fixed commandments, fixed to a people and to a time.

Theatre was a world within a world and a world where human character could be explored. Shakespeare the boy suffered devastation when his beloved sister was taken away to be buried at the age of seven. No one prayed so much and so hard as William Shakespeare, but it was all in vain. He came away knowing all about talking to thin air. Existence was hard with or without God. God failed his first test. Theatre was about creating an illusion. The young boy was the victim of an elaborate hoax and a cruel mind game. His betters sold him a lie. Hamlet is about revenge. What greater revenge could the hapless teenager have than to tell his truth to London, to England and to the world? The non-university Wit could say, "I am a man who has told you the truth." Shakespeare saw his parents lose everything they worked for. Like an avenging angel he applied himself to the task of clawing everything back.

Shakespeare was driven, he worked hard. It was to do with the resilience Mary Arden instilled in him. Shakespeare stood up to adversity. The University Wits liked him for his ambition. The upstart was assimilated. The crow was beautified with their feathers. Everyone had to survive the slings and arrows of outrageous fortune. Most had a good idea about the nature of existence, and it came from the Bible. There are some positives to the Bible.

*To everything there is a season, and a time to every purpose under the heaven:*

*A time to be born, and a time to die; a time to plant, and a time to pluck up that which is planted;*

*A time to kill, and a time to heal; a time to break down, and a time to build up;*

*A time to weep, and a time to laugh; a time to mourn, and a time to dance;*

*A time to cast away stones, and a time to gather stones together; a time to embrace, and a time to refrain from embracing;*

*A time to get, and a time to lose; a time to keep, and a time to cast away;*

*A time to rend, and a time to sew; a time to keep silence, and a time to speak;*

*A time to love, and a time to hate; a time of war, and a time of peace.*

The Roman Catholic Church was a foreign power which did well from her English assets. Church was establishment and Church made laws and Church took care of itself. It was run by ambitious career minded professionals. It was run by people wishing to serve the community. It was run by people with a spiritual motive. The Pope was the Supreme Head of this international organisation. Bishops looked to Rome for reward and promotion. As in any organisation loyalty and service mattered. Control passed from the Roman Catholic Church to the Church of England. Organisation was affected but theology remained much the same. The Church resisted Ann Boleyn and Catherine Parr. The Church resisted Elizabeth. The Church resisted any further diminishing of power and authority. The Church accepted moral responsibility for everyone, and every member of the faith family was guaranteed a place in heaven. Purgatory facilitated this.

The system was explained in Hamlet by the ghost.

*I am thy father's spirit,*
*Doomed for a certain term to walk the night*
*And for the day confined to fast in fires*
*Till the foul crimes done in my days of nature*
*Are burnt and purged away.*

Every Christian would ultimately enter Heaven which was their birth right and baptismal right. They had spiritual welfare provided from the cradle to the grave to heaven itself. They had the Church to thank. Lay people were not expected to be experts in philosophy and in control of their own destiny. Faith like propaganda needed intellectual elite to make sense of it. Ordinary people had no choice but to trust trained experts. Priests on account of being Biblical scholars played a vital role in maintaining the theocracy. Individuals seeking moral responsibility for their own thoughts and actions were a threat to the status quo. What would be the purpose of professional theologians and thinkers if the common man possessed all the theology he could ever need? The church had the way of thinking. Shakespeare introduced a new way of thinking with his new way of speaking. Poetry and blank verse made people fall in love with the English language.

The epic plays by William Shakespeare were a subtle form of social influence and social engineering. Someone was attempting to influence England's way of thinking. The plays were successful and are so good they have outlasted the Reformation society they were written for. Hamnet was buried on 11 August 1596. By 1603 there was a draft of Hamlet in existence. The play is a mix of old and new. Myth, philosophy, faith, superstition, nature, and existentialism make up the moral fabric. The prince questioned the meaning of life and gave place to conscience. The play is an existentialist work. There was an eventual church backlash and playhouses were shut down in 1642 by the Puritan brand of philosophy. A tragic loss triggered the "meaning of life" question and every time someone today repeats the line "to be, or not to be" he or she is leaning toward existentialism.

He was the oracle fast tracked into the royal courts. The Queen of England and the Church of England scrutinised his plays and gave their seal of approval. If you love Shakespeare, you love existentialism. He avoided religious persecution because the guardians of theocracy could not pretend to understand existentialism. Church and ministers of state found the plays impenetrable. One suspects the Queen was pretending not to know about Catherine Parr and the true extent of progress which she too had embarked upon. The man from Stratford had no connection with royalty or nobility. He was a crow. He came in with nothing and that was verifiable. He had a reputation for the bombastic. His money came from his plays. He belonged to a theatre company, and they attributed their success to the only one who was not a university wit and the least likely to arouse suspicion.

Hamlet is a play about revenge. We just need to stretch and squeeze the concept of revenge to see it for the existentialist mechanism that it is. We are made of the selfish gene and as night follows day, we end up aggrieved. Some are made cowards by conscience, but some are made brave. Some go seeking revenge. The God of War was vengeful. All existential creatures made of the selfish gene can say, "vengeance is mine". It is asking for trouble if everyone is made of the selfish gene but that is the way it is. Spirit beings are in denial and do not expect vengeance from God. It is not one of the fruits of the spirit. But Hamlet was not performed in a church. The principle of revenge is part and parcel of existence. We are all selfish and selfish needs to be sensible and beyond that we have laws to make us sensible. In the Garden of Gethsemane vengeance was taken off the agenda for sake of integration. Christ, Judas, the Galilean and Greek apostles all became as one.

The unwitting Judas forced God to review his moral status. Hamlet stopped himself killing the king when seeing him in prayer and thinking him to be reconciled with his maker. Revenge for his father's death was taken off Hamlet's agenda and he returned to his senses. Righteousness compelled God and Hamlet to change direction. Righteousness changed and the course of action changed with it. Hamlet was a righteous prince. It was righteous to avenge the death of his father and then it was not. Shakespeare was battered by the changing waves of fortune. He changed tack many times before arriving at the Rose Theatre. At last, he was in a place where he could take control of his life which so far had been dictated to him by circumstances out of his control, like Anne Hathaway getting pregnant. Vengeance is mine said William Shakespeare that day the first of his plays was performed. His relationship with fortune changed. He was taking control of his life.

Shakespeare's parents gambled everything on a university education for their son. They lost everything and he lost his place in school. But he had plenty education to eventually carve out a career as a writer. Academia rejected him and denied him a precious opportunity and the humiliation returned him from his elevated status. He looked for another way to become one of the beautified people. The University Wits could see the lowly player had a tiger's heart. They decided to finish his education by mentoring him. They beautified him with their feathers. In all this he never stopped being a crow. The London theatre company became his second family. Someone suggested him for their spear shaker and the plays were attributed to him. No one had heard of existentialism and so they could pass him off as an oracle with a special message for the goddess Athena. An oracle does not have to understand the message from the deity. The university educated Wits understood the message.

Shakespeare drew criticism from Robert Greene who accused him of being an "upstart". The word perfectly captures the fact that in 1592 he enjoyed a change of fortune and a change in status. The jack of all trades had climbed a career mountain and was atop a personal plateau. Greene mistook vindication for conceit. It was an outsider who was beautified, and one Wit did not think his beautification was merited. Was it ever a question of deserved? The man who left Stratford-Upon-Avon was a church-ling with a Catholic background. The fact that he wrote successful plays made him a playwright. It did not make him a University Wit. All the religious reforms, counter-reforms and reverse reforms had him in a spin. Belief was fluid. In 1592 Shakespeare was said to be bombastic. He would have sounded bombastic to anyone without appreciation of the existentialist theme running through his work. There is nothing bombastic about Hamlet, a play about personal reform.

We know Hamnet was just eleven when transferred to the undiscovered country. The wrench left Shakespeare in despair. He was driven to drink. "Not to be" was his answer to the question of existence. It was a time for depression. He had nothing to live for. Losing his son meant there was no one to carry on his name. All the beautiful feathers fell away. Naked existence was forced on him. It was forced on his wife too. Hamnet had his revenge on the living. His memory haunted his father. The family was devastated by the loss. Mary Arden and

John Shakespeare had to deal with another death in the family. But mixed in with grief there was blame being thrown around. Outrageous fortune struck young William Shakespeare when his little sister was felled. In the case of Hamnet there were contributory factors.

The absent father was obliged to spend a lot of time in London, for which he was richly compensated. The family prized him for his regular monetary contributions. But when Hamnet died their love for him turned sour. It turned into dispriz'd love. No amount of money could make up for his absences and the awful tragedy. The family was in pain. The only way for Shakespeare to deal with his feelings was to get it all out in the open. He was a writer and he needed to write it all down. He needed to express himself if only for healing's sake. He ended up writing another epic play. Shared existence underlined the precious opportunity for a personal relationship with someone. When robbed of Hamnet's existence all he had left were memories. It was in the spirit of Shakespeare to share the life lesson. Existence was something he had to share. It might help someone.

The play was the thing. The story began with a ghost but by end the ghost was gone and naked existence was experienced and then came death and finally all that was left was a memory. It is no coincidence that "Shakespeare: Rise of a Genius" chose for its first Shakespearean quote, "I've taken too little care of this." The BBC explained that his commitment to his family suffered because of his commitment to his writing. The plays demanded the best of him and as part of that commitment he ended up spending a lot of time in London. There were limits to being William Shakespeare. He could not be in two places at the same time. Real life inspired him to personalise the story of King Lear, a play about taking too little care of family. He was borrowing from history and experience and creating a depth to self. The self was not just about suffering injury and wishing ill on those responsible or even revenge. Yes, we are reactionary but whatever emotions define us they colour our souls. It is character we need to take care of.

Protestantism began with Martin Luther in 1517. He was kind enough to write it all down. The points he raised were answered by a staunch Catholic, it was Henry VIII. The Pope was so impressed he honoured the Catholic king with the fabulous title "Defender of the Faith". That was the highlight of 1521 for the 30-year-old Catholic king. His name was entered into the pages of church history. Henry was in a state of bliss, and he wanted nothing more than to live up to the holy title. In 1534 the Church of England was formed. The Catholic king still believed the doctrines he had earlier defended but found it necessary to re-structure the organisation and transfer control from Rome to London. It was another kind of reformation. Everyone in England held the same beliefs they had in 1533. The king divorced and remarried women with varying degrees of Protestant sympathies. The Reformation was a bigger problem on the continent where the 2 sides were warring.

The Defender of the Faith never tired of his grand title. He had advisers to help him keep his place in church history. The Devil had him in his sights. The Devil was known to be cunning. He was the invisible enemy who could use his friends against him. Spiritual warfare demanded special attention. The war was between faith and lies or propaganda and lies. Into this cultural quagmire stepped the first existentialist since the establishment of Roman Catholicism and the Holy Roman Empire. It was the beautiful soul known as Catherine Parr. She was the first person since 325 to enter a personal relationship with God through Jesus Christ. Her personal relationship was with God directly. It was not with God via pope or church or any other Christ substitute. Indulgences were symptomatic of a much deeper problem which went to the core of "church". The very idea of church was brought into disrepute.

When Jesus Christ is not being Jesus Christ, he is God. Catherine's personal reformation made it incumbent on the church to make clear that it was not placing itself between Christ and his followers. The role of intermediary needed to be defined. Christ said his kingdom was not of this world, yet church took it upon itself to make an empire of itself. There ought to be a Christ kingdom, just not of this world. Christ was a personality and not a political system. Philosophy has a creator-God which is a God in principle. Jesus of Nazareth was not God in principle. Theocracy is not for Christians because they belong in a secular society. Christian principles are bound to filter through, and representation is necessary. The Church of England still had an important role to play in the life of Christians but not in the life of the nation. No one wanted the Church of England destroyed. They wanted it functioning for Christ and not in place of Christ.

Meanwhile, those bishops functioning in the royal courts were not so much Christian existentialists as Christian philosophers. Catherine's beliefs were perceived to be a threat to their church and an attack on their faith. A death warrant was issued by her husband and advisors. The Defender of Faith was Henry's get-out-of-Purgatory card and without it he would have to face greater torment. The much-loved Catherine had the royal household divided. Her death warrant ended up in her own hands and in the summer of 1546, she was forced to make a life-or-death decision. Some Christians will choose heroism and martyrdom. Some will leave it in God's hands. But Catherine was desperate for the truth to be known. She needed to know the full truth and for that she needed to survive the latest purge in the royal household. She was spared the axe with a vow to Henry that she only argued about religion to take his mind off the suffering caused by his ulcerous leg. She didn't mean any of it.

The maxim "to be or not to be" is her story condensed to one line. The Queen was England's first existentialist. She too could be as duplicitous as the Defender of Faith and Supreme Head of the Church of England. "To be", that was her answer. From then on, she knew better than to discuss religion with the King. The details were sure to end up with the bishops and they did not believe in free speech. Freedom of speech became an issue and Catherine more than anyone knew to be mindful of her words when speaking to powerful bishops and members of the powerful Privy Council. Catherine passed on all knowledge of court protocol to Princess Elizabeth who was her stepdaughter. Elizabeth's own mother, Ann Boleyn, was murdered by the same forces that threatened to murder her stepmother. The young Elizabeth was getting to be quite the expert on church and court methodology.

Elizabeth was bastardised till Catherine Parr intervened on her behalf in 1543. In 1558 and at 25 years of age Elizabeth became Queen. She was determined it was time for the women to fight back. She too had spent time locked up in the Tower of London. She was groomed and manipulated onto the edge of treason but that was 1554. Aspiring bishops and ambitious nobles were always jostling for power and influence. Elizabeth knew what it meant to be used by an immoral man and persecuted by the system. Only the Queen could take the fight to male chauvinists and misogynists in and around the circles of power. It was a shock for the men to be resisted and outwitted by the intelligent, confident, and fiercely independent young Queen.

The Virgin Queen was married to a cause which demanded more years of her every year she was married to the cause. It was time for revenge. Revenge had to be carefully planned out because it had to be served cold. Revenge most cold, as in the best it is.

# Chapter 3

# The Reformation in Plays

In the summer of 1546, a death warrant for the Queen of England was issued. It was intercepted in the gardens of Hampton Court Palace by the Queen herself. Catherine Parr died upon reading the judicial order for the death of another queen of England, namely herself. The warrant stripped her of regalia and privileges and prerogative. Her neck was on the block and the warrant was the axe bearing down on her. She was brought face to face with naked existence. She was without hope and her rationale for existing or reason for being was torn away from her. She was to be ex-communicated and condemned to die. She was truly on her own against the Church of England, the religion of Europe and the conspiracy of philosophers. But that licence to kill gave rise to existentialism.

Interception of the court order gave Catherine prior notice of the martyrdom planned for her. She would have days in the Tower of London to reflect on her heresy. She was a religious person and bound to conclude integrity was worth dying for. She would eventually be resigned to her fate. After many offers of confession and repentance for the sake of her eternal soul Catherine could be predicted to accept martyrdom. But for her it was a real choice and Catherine Parr decided to fight for her life. She was willing to give up her personal faith and like Desdemona beg for one day at a time. She changed the lens from faith to existentialism and wisdom increased exponentially. She could live to fight another day. Granted existence she could reveal the truth and avenge the injustice later. She was an existentialist looking for vindication. She was the guardian of Princess Elizabeth.

For a good part of the play Hamlet's existence is for the sole purpose of revenge. The advantages of being a royal and future happiness with Ophelia could not compensate for his pain. Revenge is sweet and satisfying and he needed it desperately. "To be, or not to be" or raw existence ends up paired with revenge. There is a lot of revenge in the First Folio. Richard Dawkins wrote "The Selfish Gene" and in it he revealed how we the product of genes learn to distinguish between selfish and optimal-selfish. Another way of looking at morality is optimal selfishness. Yahweh-Christ was a selfish creature. Not meaning selfish as in acquisition but selfish in that he was about the self, his own self. Christ was about his own emotional welfare. Yahweh-Christ was also a social creature. Existentialists want a fulfilled self. Existentialism thrives in a meritocracy. Christ the emotional powder keg was a Christ despite theocracy. He was not a theocrat.

St Paul who is the champion of philosopher Christians had something to say about existential Christians.

*For the invisible things of him from the creation of the world are clearly seen, being understood by the things that are made, even his eternal power and Godhead; so that they are without excuse:*

*Because that, when they knew God, they glorified him not as God, neither were thankful; but became vain in their imaginations, and their foolish heart was darkened.*

*Professing themselves to be wise, they became fools,*

*And changed the glory of the uncorruptible God into an image made like to corruptible man.*

St. Paul was a champion philosopher, and the God of philosophy was his God, and he is the apostle in Apostolic Creed because he was able to bury the truth with his treacherous "Christ-after-the-flesh". Christians, you know, reject Christ for philosophy to this day. They get taken in and spewed out.

Catherine Parr's revenge was served cold because revenge was by the hand of William Shakespeare. Philosophers wanted a benevolent creator God because it led to civilisation which was their thing. It was a time before democracy and republics. Philosophers did not want a self-centred God who came with emotional baggage and with no interest in the affairs of men. They managed to keep buried the God being dug up by readers of gospels in the vernacular. They were very good at hiding the truth till the printing press came along in the 1540s. There was a lot more philosophy around before the German New Testament of 1522, courtesy of Martin Luther. The King James Version was first published in 1611 and by then consciousness of the existential had increased exponentially. The Tyndale Bible is of interest because that translation into English was completed in 1535. There was no power like the power of the written word and Shakespeare was a genius at it.

Indulgences was not the first faith crime against the people of Germany. It was just the first which could not be successfully hidden behind the vails of philosophy. The people of Germany were victims of church propaganda. (I joined a church for the social life but soon found myself warring against the devil. So much of my valuable time was wasted fighting something that does not exist. I am now at war with the church of sabotage.) The church accepted moral responsibility for her devotees. The Russian Orthodox Church accepts moral responsibility for naive conscripts fighting in the 2022 war against Nazis. As in any war, truth is the first casualty. In the existential universe demonic possession is mental illness and the Nazis were defeated in 1945. Existentialism is like science in that it identifies human attributes and conditions which neither church nor the supreme head of church can change. Without the existential truth you end up getting rampant spirituality.

Luther had to deal with greed and corruption on part of church professionals. Catherine Parr had to deal with murder and duplicity. Science verses religion is another story from a later time. In the Elizabethan era it was history stolen by philosophers and Christ wilfully misinterpreted by self-serving Roman oligarchs. Reformers were emboldened by literacy. We first see the selfish gene in operation in plays like Hamlet. Elizabeth did not write plays although someone with her education, background, and able to speak five languages could have given it a go. She gave England one of her greatest ever speeches and the Spanish Armada was vanquished. She must have written a soliloquy or two in her time. But Elizabeth could authorise the existentialism of Catherine Parr. Her guardian was a renowned educator and although she was dead her teachings lived on.

Elizabeth avenged the death of her mother who was taught to be accepting of death by the same church which murdered her. She avenged the terror inflicted on her stepmother. Her personal revenge took till 1623 to manifest. It was the end of philosophy and the male chauvinist philosopher. Elizabeth knew the enemy well enough to play out her role of avenger without setting off warning shots. That did not stop her getting ex-communicated by the Pope and the Roman Catholic Church. It made her a target for the Spanish. The church was and still is the greatest threat to humanity. They excuse their crimes on basis of church unity. Elizabeth refused to give them England. Elizabeth wrestled philosophers for the freedoms England enjoyed and which before her time were non-existent.

She was not afraid to exercise her divine prerogative against her bishops. Freedom for Elizabeth meant not being coerced by either the Roman Catholic Church or the Church of England.

Elizabeth's future England first appeared on stage. It was a revolution. It was a mix of philosophy (Catholic and Protestant), fantasy, and existentialism. It was not theocracy. It was not a return to the world as it was before the Reformation. The new England had a good idea of self. It was open about human character. It was about the selfish gene, jealousy, and revenge. It was about morals. It was about having to put up with philosophy without losing sight of the existential. England would have to wait till 2006 before it could be said that philosophy is delusion. Shakespeare put his name to a new way of thinking. He was bent on providing his audience a new raison d'être. Theatre encouraged people to think for themselves. They were not in church, and it was not Sunday. People were free to judge the fictional characters and dip their toes in the sea of moral responsibility. Every revision of every play was moral evolution. Hamlet after many revisions is the Elizabethan flagship of existentialism.

Plays gave thought multiple roles. Thought came in guise of truthfulness, superstition, and deception. Action had to resort to scene stealing. William Shakespeare was the name on the front cover because he was a political non-entity. He was the best candidate to deflect scrutiny and suspicion. Selective Retrospective Existentialism through church records confirm he married Ann Hathaway, and they had three children. We know where they lived. We know what William Shakespeare looked like. We have his image in 2D and 3D. We know he was born in Stratford-upon-Avon. We know a narrative poem and a book of plays was published in his name. We have no record of anyone else claiming authorship. The Globe theatre did not dispute authorship of the plays, they were the ones who assigned authorship to one William Shakespeare. Their project and reliance on him ended in 1623.

The plays were a different matter. They continued in need of protection. The Globe decided the name of William Shakespeare was still the best shield for the literary masterpieces. His association with the Holy Trinity Church gave him a religious background and that was good reason for the faith family to leave well alone. Success for the stagehand resulted in him being called an upstart which he was. There were no fine feathers on him. It was a case of what you see is what you get and what you got was a crow. But he was ambitious and doing his best to climb the social ladder. He was the humble stagehand who made a success of himself. The rise of a genius was originally the rise of an upstart. Other writers saw in him a magpie. But audiences were slayed by the catchy new soundbites. How wonderful to have penned, "to be, or not to be, that is the question. The Wits included him in their company. He had all the help he needed. He had the full support of everyone listed in the preface to the First Folio.

The upstart crow climbed his way to part owner of the theatre. He was fully assimilated into the theatre fraternity. The Wits made it their business to beautify the crow with their feathers. The actors and writers, some of whom were big names, made conspicuous their appreciation of his talent. The survival of the theatre and the plays depended on him. They were safe to indulge in their art form because nobody felt threatened by the Catholic worker made good. The BBC pointed out "upstart" was a new phrase in 1592 because for the first time in history a man could change his social status. Theatre achieved meritocracy and a jack of all trades found his career-calling which was to be a writer. Upstart crow or no, the theatre group recognised his talent and rewarded him suitably.

The state was always in need of affirmation. Speaking out was not without consequence. Always affirmation although sometimes challenging to reflect the times. Theatre put on spectacles and operated to commercial interests. Any political message had to be hidden deep in the play. The work was anyway quality controlled by their patron the Lord Chamberlain. The Queen could command a performance. Theatre was serious endeavour. There was a building dedicated to the art form and it drew creatives to the new profession of drama. The management could afford to be choosey when hiring talent. The University Wits was a new intellectual elite, and members were vetted. From the beginning the entire organisation was about the plays. It was all about the plays. The First Folio says so. If after all the attention to detail the Queen's ministers had an objection, she was first to know about it.

Shakespeare was able to write 37 plays during the Reformation and as part of it. England and Spain plotted against each other, but we remember the plots to Romeo and Juliette, Hamlet, Othello, and other masterpieces. The aspiring businessman with no connection to people in power was left to carry on making his money. He managed not to alarm the people in power. He was an unknown quantity, and this threw off his critics. When he penned "I hate the Moor" no one asked why because everyone hated the Moor.

But we know Shakespeare's streams of consciousness were neither bombastic nor random. They took root and changed civilisation forever. Theocracy was injected with small doses of realism and the cult of the self. The philosophical elite were dismissive of everyone else. Philosophy went back to Graeco-Roman times. Comparatively speaking Shakespeare was a fad. Rumour had it that the glover's son was an oracle channelling the thoughts of Athena, goddess of the arts.

Philosophy swept up Jesus Christ and used him to consolidate civilisation as it was known to philosophers. Philosophy swept up science by claiming it to be Intelligent Design. Philosophers similarly hijacked cultural and social advances, including Shakespeare's. By their conspiracy all good things come from philosophy, ultimately. A modern example of philosophers deceiving themselves and everyone else is the propaganda projected by the Russian Orthodox Church. If they win the war against Eastern Europe, they will build another cathedral and re-write history and the worshippers will be none the wiser. They have already forgotten the Soviet Union which tore down cathedrals. They glory in philosophy and every bout of starvation experienced between glories was a test of faith. Orthodox, Catholic and Protestant Christians live lives dedicated to the glory of philosophy.

The Reformation was an attempt to break away from philosophy because it was attempted by people who believed there are more things in heaven and earth than are dreamt of in philosophy. No one can understand propaganda because it is not meant to be understood. Faith and philosophy are not meant to be understood. A Jesus who is bigger than the galaxy is not something anyone understands. Social engineering for the sake of a philosophical civilisation is what philosophy is all about. It was not possible for me to make a categorical rejection of divine omnipresence till 2006. It was not possible at all for someone living 400 years ago. Philosophers assumed everyone believed in divine omnipresence. Reformers did their best to untangle themselves from the mess. Shakespeare succeeded in preserving something of the truth because his plays were existential. They will remain relevant for as long as the human species remains an existentialist one.

A philosopher was someone with encyclopaedic knowledge because he had a theory of everything. Everything was based on Intelligent Design which itself suggested there was a designer phenomenon. It was part science and part theory. Design was an attribute applied to the scientific, the social and the spiritual. Therefore, everything was part of the cosmological argument for the existence of God. Universe and civilisation belonged to the philosopher. Ancient philosophy invented the creator-God who put design into everything. In the 16th Century philosophy kept being challenged. Catholicism was a brand of philosophy. Theology and spirituality came out of philosophy. Scribes, Pharisees, and prophets crawled out of philosophy. When reality does not conform with philosophy it is because of an existential discrepancy. Mary, Queen of Scots was ambassador of the Catholic Church. Her son James VI was a protestant. In 1585 he rejected his mother's plea to rule Scotland together.

The Reformation left one major scar on the thoughtful and feeling queen. Having avoided the bloodshed seen in France she was forced into doing the unthinkable. In her mid-fifties the daughter of Ann Boleyn reluctantly ordered the judicial death of a queen. There was need for greater empathy than that shown by the royal advisors and manipulators of the day. As if an answer to prayer we find 5 years down the road a certain upstart making his name. His name preceded the plays. Were it not for the name William Shakespeare and his humble origins the plays might have been too risky to publish in that divisive age of plots and gunpowder. His name is the reason why it had to be him to write the plays. Shakespeare is a name to inspire revolution, and it inspired free thinkers to act against social control and social engineering by philosophers. Philosophers had their own language which they called the Word of God. Shakespeare had his own existentialist language and if it was the word of anybody it was that of Catherine Parr.

It was one thing to publish under a false name but something entirely different to publish in the name of another living person. They kept him for the name and upgraded the person to a Wit. William Shakespeare was reinvented and set up to succeed. He agreed to his enrichment, he positively encouraged it. It probably started as a workplace joke. The new stagehand was introduced to his fellow workers. He was from Stratford, and he spoke with a country accent. Pronunciation invited ribbing. Someone decided to invert Shakespeare to spear-shaker. The name stuck because there was a certain irony in having a "spear-shaker" working in the arts. Athena, goddess of the arts, is usually depicted holding a spear. She is also known as Pallas-Athena. Pallas means to shake. The University Wits obviously liked the connection. But there was more. The name William is derived from the German Wilhelm which means "helmet of". Helmet and spear-shaker made for an irresistible image of Athena.

Athena wearing a war helmet and spear in hand is the image of battle readiness. It is preparedness to the point of giving the order to attack. The decision to go to war is not taken lightly, not even by a goddess. One can assume the goddess had been provoked into war and is having to go to war to impose her will on those defying her. The deity needed conviction to act. All other avenues had to be exhausted first. She needed just reason for her fury and for entering armed combat. A deity is reactionary and does not live in a permanent state of exasperation. Athena armed with spear was only because there was no other recourse. It was a time for getting angry. Anger was validation. It was a time for war and there was a way of knowing it was the time. Her conviction and rage worked up was the cause of shaking. Her arm would start shaking and the spear would start shaking. It was her force of will causing involuntary shaking.

The symbolism was clear. The direction was clear. It was tempting to think providence was on their side. The instruction for battle to commence came in the form of a name. The theatre company was emboldened. Good fortune gifted the freethinkers of England a spear shaker. "William Shakespeare" started a war. His appearance signalled it was time for action. The Wits had ink, quill, paper, an art form, and a dedicated theatre building. They knew how to put on a show. They had artistic flair. They were in a state of readiness. It needed an act of will on the part of the Wits to instigate revolution through drama. Will shake spear was their rallying cry and they did. Have the will and shake the spear come what may. It was another way of protesting against the lies and delusion which made up philosophy. The best part was that he was a Catholic from Stratford-Upon-Avon.

The genius of William Shakespeare was that he was a real person. There is a question of authorship over the plays but that came later. While he was alive, the man from the theatre co-operative only needed to pass himself off as William Shakespeare. He was made to look like a Wit, and he was always in the company of Wits and that was enough to pass him off as a Wit. He was also the upstart crow. His "crow" behaviour and other inadequacies like his rough manners meant he was not the ideal after dinner speaker. Upstart crow was the cover. The actors were the stars everyone wanted to meet, and they sold Shakespeare the playwright to their adoring public. Hamlet is the perfect example of a play inspired by events in Shakespeare's life. From the strained relationship with his daughters came King Lear. The Stratford man had Christian sympathies and his philosophy had to be incorporated. The plays introduced Christian existentialism.

Catholicism is a brand of philosophy which was established in 325 and in guise of the Roman Catholic Church lasted into the 16$^{th}$ Century mostly without challenge. In 1546 Catherine Parr became the first Christian existentialist. She refused to die for her beliefs. "Know Thyself" requires living and is therefore a mantra for existential beings. She kept herself alive rather than play the game of life by the rules of those trying to dispose of her. The world is a stage but off stage the source of Elizabethan existentialism was Elizabeth herself who in her impressionable years was mentored by England's finest educator. The Queen employed Edward de Vere to project manage.

Two other notables were John Heminges and Henry Condell. There was no question of authorship till authorship was questioned and by which time it was too late. England acquired existentialism for Everyman and by William Shakespeare their Everyman. Philosophy gave England concepts of cosmos and civilisation.

If we time travel to Shakespeare's time we lose all trace of evolutionary biology. We find Apostolic Christianity and Adam and Eve defining what it means to be human. Biology and theology both deal with dualism but when all you have is supernatural dualism you do not know there is another way of explaining the phenomenon of conscience. The Word of God has supernatural dualism because it was the only explanation for dualism that existed before the notion of innate dualism. Ancient philosophy explained conscience or dualism within a scheme of spirituality. They were sure natural life was supplemented with supernatural life, an external lifeforce. Death brought natural life to an end but supernatural life continued. Natural life is existential and the supernatural was thought to be eternal. Philosophy has a twisted outlook on life and existence. In short, they invented a spirit universe to go with their supernatural human being.

Hamlet's comment on philosophy does linger in the mind. "There are more things in Heaven and Earth, Horatio, than are dreamt of in our philosophy." There is also the selfish gene, evolutionary biology, Big Bang, and electric cars with GPS. If it was Stephen Hawking's play he would have said, "Horatio, philosophy is dead." Hamlet Senior speaking from the grave is indicative of supernatural dualism. He said of purgatory, "till the foul crimes done in my days of nature". The writer distinguished nature from the philosophical mechanism known as Purgatory. Shakespeare challenged the assumption that philosophy is the only operating system for the universe. The playwright lacked science and technology we take for granted.

He could not make a categorial rejection of philosophy. He said what we all know, the idea that philosophers know everything is an illusion. They are experts at hiding the fact that they do not know anything for a fact.

The Holy Trinity Church of Stratford-Upon-Avon kindly recorded Shakespeare's date of baptism. The townsfolk paid tithes and the church provided civic services. The holy trinity and the Apostolic Creed are the theology of St. Paul. No theology can be attributed to the Galilean apostles who were unlearned men. They were men chosen for their ethnicity. Paul was full of faith. Christians are faithful to a fault. Paul was an ethnic Jew like the Galilean apostles, but he was also a spiritual Jew. He was a Pharisee who was conceited about his spiritual ethnicity. He was born a Pharisee. The Galileans could not make any such boast. Fishermen is not a spiritual status. In the confused world of spirit and flesh Pharisee was as much an ethnic status as Jew. Christian was Shakespeare's spiritual ethnicity. Hamlet senior was Christian by ethnicity, and he found himself in Purgatory paying for his crimes. We are told more about the operating system by the playwright.

*Doomed for a certain term to walk the night*
*And for the day confined to fast in fires*
*Till the foul crimes done in my days of nature*
*Are burnt and purged away. But that I am forbid*
*To tell the secrets of my prison house,*
*I could a tale unfold whose lightest word*
*Would harrow up thy soul, freeze thy young blood,*
*Make thy two eyes, like stars, start from their*
*spheres,*
*Thy knotted and combinèd locks to part,*
*And each particular hair to stand an end,*
*Like quills upon the fearful porpentine.*
*But this eternal blazon must not be*
*To ears of flesh and blood.*

Cosmos and theology equate to universe and operating system. Christianity was a model of the universe. A working man had to rely on priests and church to get it right. Preaching material could be educational because of "intelligent design". The natural world was a place to find God's handiwork. History which went back to 325 and beyond to the apostles and even to creation was reassuring. The system tried to offer the widest experience of life. The system worked better for some than others. Fear of purgatory and fear generally played a negative part in the psychology and mental wellbeing of God's people. Then came the printing press. People were shocked to find that almost nothing they had been taught came from the gospels. There was a big downside to books and literacy and the freedom to discover Christ's love for themselves. Christ was not about the system. Christ was a humanitarian and stayed within the remit of a divine personality cult.

People learned to distinguish between Christ and philosophy. Once again, philosophy explained away negativity as sin. Socrates invented sin to explain away Athens's apathy for his wild speculations. The Protestants were guilty of sin on a grand scale. The difference was that in a theocracy it could be punished. In Europe sin and anti-Christ papacy separated Catholics and Protestants. In England, thanks to Elizabeth progress was allowed to take its course. Philosophers used their expertise at hiding the truth to bury the reformation. They kept it out of their schools and churches but could not keep it out of theatres. Philosophy is an intellectual discipline and before you know it you want to keep out anything that challenges it. Yahweh the God of War was to philosophy a myth and a superstition. He was a joke. Christ the Logos underpinned the system. The University Wits and free thinkers wanted reason to prevail and science freed.

Catherine Parr was another well-read person and she looked to philosophy for answers. She debated religion and she talked

Christianity. She too expected faith to yield better results. Instead, the defenders of faith turned against her with a death sentence. (I gave up trying to make sense of faith in 2006 because I was by then ready for Richard Dawkins and the total abandonment of faith. I discovered real-time and prophecy was dead. Faith at a certain point became a hindrance to my progress and personal development.) The Queen discovered just how toxic was faith because it was inconsistent with the world outside of philosophy houses. The church wanted to strip her naked and terminate her existence. They managed to strip her to her soul. She was born again an existentialist. The truth did not come out of Stratford-Upon-Avon nor Oxford. It came out of Hampton Court Palace.

Revenge flowed from Catherine through Elizabeth and Edward de Vere to the University Wits and William Shakespeare. It was served cold. People started thinking for themselves. That was the revenge. It was revealed 77 years later courtesy of John Heminges and Henry Condell. The project with all revisions completed could be viewed in entirety for the first time in 1623. All the key players and Anne Hathaway were dead. Hamlet revealed conscience as the guide through his sea of troubles. He always did what he thought was the right thing. It was as much as any theatre audience could reasonably expect. He was righteous. There was no confession to a priest for his wicked thoughts. Philosophy is subject to existential discrepancies and when that happens let conscience be your guide. The existentialist plays were impenetrable for those with a philosophical mind. The Reformation was division of conscience and faith.

Spear shaking is a lot quieter than sabre rattling but not dissimilar. Intent was put on display. The enemy was to be made aware of the anger. Those occupying the centre ground were not defeated but circumvented. In 1623 the First Folio institutionalised spear shaking. It came with a subliminal instruction on the cover to shake spear. Philosophy had started off as the pursuit of truth but lost its way. It became about theocracy which like any constitution involves social engineering. Russian oligarchs came into existence and soon enough they too were about social engineering and saying things like, "I can make a monkey president." Philosophers want control. All the good philosophers died a long time ago. Philosophers of today waste money and human resources fighting sin, the flesh, and the devil. Philosophers send young men to their deaths and glory in the fact that they blessed them.

Social engineering cannot be avoided. The Kremlin is best at it. The patriotic war was invented for peace time. It covered a multitude of sins on the part of Stalin and his cronies. The patriotic war has crossed over into a real war, but it continues to cover sins, the multitude of sins on part of the imitation Stalin. Russians like a strong leader. The supremacist engineered circumstances so heart-breaking that war was the only source of happiness left to him. In the 16th Century the church in Rome assumed moral responsibility and issued diktats for Christians in England to abide by. Catholics were never happier. Martin Luther brought transparency to church methodology. Tithing was a reasonable demand, but the new methods of revenue generation were nothing short of cynical. There was a lot more wealth and the church insisted on increasing her lot. The sale of indulgences was a way of fleecing her parishioners.

The fear of Purgatory slowed down sinning a lot more quickly than conscience. But conscience makes for character. The system was put under scrutiny. Spirituality is the art of denial, and centuries later it was about the denial of science. In the 16th Century spirituality was about the denial of history. History was the framework for understanding Jesus Christ. Spirituality was for academics and the intellectual elite. Christ was a living breathing person and what was known of him was being suppressed by the philosophers. The only thing existentialist hate more than a person who refuses to acknowledge history is the person who revises history. Worst of all is the person who destroys history. Therefore, the greatest existentialist statement ever made was put into the mouth of Iago who said, “I hate the Moor”.

# Chapter 4

# I Hate the Moor

But how to fight the Moor? The existentialist appeals to reason while the philosopher Moor turns to intimidation and violence to get his way. He has it in his mind that as a man with divine revelation he is to be honoured. God has honoured him and so must everyone else. It is not a human right but a spiritual right. It is a divine right, and he demands everyone yield to the divine will. Existentialists are apostates and enemies of philosophy. There is no disguising the hate and ill will coming from both sides. Anonymity is the best form of defence when debating with any member of the insidious faith family. Every member of the faith family is a defender of faith, and they are all looking to make a bold statement of faith. Socrates was first to employ faith for the purpose of social engineering and he died to demonstrate his faith. Throughout the ages many have followed his example.

The apostle Paul wished someone dead for the sake of his philosophy. Stoning used to be a favourite hobby of his in Jerusalem. He said about a Corinthian, *To deliver such an one unto Satan for the destruction of the flesh, that the spirit may be saved in the day of the Lord Jesus*. The faith family reserve the right to silence anybody who speaks out against their prophets and their philosophy. Anonymity frustrates them because they are muted if they have no one to charge with blasphemy. There were outbreaks of religious fervour in her kingdom, but the Queen was able to keep England on an even keel. They all needed time to reflect on the changes being proposed by the Reformation. It was Elizabeth who permitted thought-provoking plays in her realm. In one play, consciousness was increased by instance of a Moor. In Shakespeare's work there appeared a Moor with Christian pretences. The Moor was known for his brutal conquests and policy of forcible conversion. His manifesto came out of philosophy. He too called it revelation theology.

When Christian Europe first heard of the Moor the response was out-and-out hatred. They had taken Jerusalem. Empire building and assimilation included Jerusalem. The response of Christians was to intervene and take back that city with which there was such a strong historical connection. Jerusalem was the birthplace of Christianity. It was where God declared he was Jesus Christ. Ineptitude on part of the Moor led to God being completely wiped out of history on his side of the fence. Retrospective application of his brand of philosophy to the God of Abraham was not without consequence. Christianity, the rival philosophy, was brought into disrepute. The legitimacy of the Roman Catholic Church was questioned. War proved to be futile because neither side could win. The Moor did more than make Jerusalem his trophy. He made a mortal enemy of every Christian and every existentialist.

Divinity, Intelligent Design, cosmos, and civilisation were ideas which game out of Ancient Greece. Aristotle was the biggest influence. Sadducees were Aristotelean and Pharisees were Socratic. Christianity was a pick and mix bag of ideas from the Greek market stall. The Moor came away with his own bag of ideas. According to the latest mishmash the Christ did not die on the cross. No death meant no forgiveness of sins. No forgiveness of sins undermined the scheme of divinity which made a god of the Christ. Without divine purpose there was no reason for assuming messiah was born divine or that divinity was merited and assigned. The socio-spiritual operating system of the Moor was very different. Ordinary people beat their ploughshares into swords and made the arduous journey across Europe into Asia for the pleasure of killing a Moor, a new age philosopher from a distant place.

We must accept the fact that hate belongs on the moral compass. We cannot deny hate on basis that we are philosophical beings because we are not. Hate needs to be processed and forgiving is an option. Due process is another. As moral beings, we need to process hate to avoid the self-loathing which often takes its place. We need to know who is acting hatefully toward us to avoid self-blame. The creator God does not hate because it is not an attribute of the spirit which is what it is. The philosopher acts hatefully when hiding or omitting the truth. We are saved from years of research and learning by transferring moral responsibility to the social engineers. How many Christians end up ex-Christians and hating Christ for what the church has done to them? They end up denying Christ ever existed. Philosophers hate existentialists and existentialists are equally committed to hating philosophers. Iago said it best, "I hate the Moor".

The fruits of the spirit are of no use to fruit pickers and moral beings. Spirituality is not morality. Spirituality and morality have been blended. Before the Reformation there was a moral void in the houses of worship run by theologians. Many Catholic beliefs and traditions needed to be modernised and the Counter-Reformation took care of it. Pope Pius V abolished the sale of indulgences in 1567. Yes, the same tyrant who ex-communicated Queen Elizabeth I. Philosophers simply appropriated the best Protestant ideas because it stopped people leaving the church. There was a very real danger of philosophy swallowing up Protestantism in the same way it swallowed up history. In the same way the church today swallows up the Big Bang Theory of 1929.

More and more people know they are being deceived. The system is designed to deceive itself. We know it is hatred of history when someone chooses not to acknowledge Christ-after-the-flesh. The faith engineers have a job to do and that is all they care about. They are engineers building a theocracy and have no interest in realism and the realistic. Long term employment is the reason why we have a Jesus who is bigger than the galaxy. "I hate the Moor" is spear shaking at its most violent. It is Iago verses Othello and the audience not only witness Iago's revenge but must pass judgement on him. They must take a side. Iago is trapped in a world full of philosophers. He is not given the lines to deconstruct philosophy of either the Moor or the Christians of Venice. It is enough for Iago to say he hates the Moor, and he does so even at risk of being repetitive. One gets the feeling that Londoners were delighted to see a philosopher get his comeuppance.

From Iago the message is very simple. Philosophy must be halted at all costs. Shakespeare's audience knew church law. They knew blasphemy would not be tolerated. No one was free to say I hate the philosopher. The starting position with Othello is hate. It must be hatred because of his crimes against humanity and the new humanitarian Christianity of the Protestants. The Moor was a backward step in civilisation and a return to philosophical totalitarianism. Protestantism was beneath him. Paul the philosopher overruled history because he did not like what Christ said or the company he was keeping. The Moor overruled history because his narrative demanded a Christ without the attribute of omnipresence. He was a monotheist. He reinvented Christ to go with his philosophy. "I hate the Moor" defines Iago. "I hate the Moor" defines the play. The hate is the thing, and it is the backdrop against which everything unfolds. Honest Iago is honest by the revelation that he hates the philosopher.

Iago is the villain on stage, but he is not the villain in the theatre. He is a complex character who enjoys honesty when he can. He can play at honest and honest Iago is a name which suits him despite the dishonesty and deception. He can afford to be honest with those who do not jeopardise his revenge plot. He quickly establishes there are two sides to him. "I am not what I am" signalled to the audience they needed to look beneath the surface and even outside the play for his motivation. In the play Iago was looked over for promotion. When asked to explain himself Iago excused himself with, "Demand me not". Shakespeare excused himself because he knew he was subject to censorship and punishment by philosopher bishops. England did not have freedom of speech. She had freedom of faith. In the play Venice has a marriage of convenience with the Moor and Shakespeare might have been pre-empting a similar pact involving London. For Christians the villain is the Moor because to be a Moor is villainous. The proof is in the murder of sweet Desdemona proving a leopard cannot change his spots.

Iago the soldier won battles and the trust of his lord. He honoured his master by putting his own life at risk for him. When it came to expressions of loyalty, he was not slow in coming forward. The unsuspecting Moor insisted on honouring the loyal Iago in some way and titled him Honest Iago. But he was honest for a sinister reason. Iago was obsessed with Othello's welfare because he wanted to be the cause of death. The pleasure was to be his. His was a detailed vengeance played out. There was a philosopher on stage. The deplorable and deluded wretch thought himself honourable. Iago must have been an existentialist because the Moor is beneath contempt. Iago did not think the philosopher deserved an answer to his question, *why hath (he) thus ensnared my soul and body?* Iago was just about one thing, delivering judgement on the Moor. He told the Moor, *from this time forth I never will speak word.* Iago did not believe in casting pearls to swine.

The funny thing about the Moor was that he could pretend to be a Christian whilst at the same time rejecting everything about Christ. He managed to fool Venice but not Iago who gave himself the extra duty of punishing the liar and fool for his deceit. The philosopher was all about deceiving himself with the notion of spiritual supremacy and spiritual arrogance. He had discovered divinity for himself. Iago was all about shattering the delusion and bringing him back down to earth. Iago took it upon himself to goad the Moor into revealing his true nature. It was a warning to England. The audience is left to decide if the maltreatment of the Moor was excessive or necessary to bring out the truth. The Moor was not someone who anticipated a spell in Purgatory. His defence for murder was his honour, the honour which comes from gnosis. He was without conscience. He was a savage beast on the prowl for ideologies to devour. He devoured Desdemona.

Iago never saw Othello as the Moor of Venice. He only saw a Moor in Venice. He was a Moor and Venice was desperate for his services. Iago was there to remind the audience they were looking at a chameleon. It was the same Moor who was originally answered with the Crusades. He had conveniently evolved into a Christian. Deception meant to hide his destruction of history. The Moor's recent past in world history could be traced to an eternal God and he accepted it as his commission to establish eternal theocracy and eternal civilisation on behalf of his eternal God. Christians were doing the same but for longer. The Moor embodied philosophy and he was the target in its most convenient form. England did not need an introduction to the Moor. Shakespeare was saved that task because although the brand was new philosophy was not. The best way to process hate of philosophy is to have nothing to do with it. That was the instruction given to Apollos by Christ. The second-best way is to follow Iago's example, expose and eliminate.

Career philosophers are professional liars. They say the God of War does not exist. They say Jesus is as big as the solar system. They say Jesus Christ did not die on the cross. They say man is part biological and part spirit. You do not have to be an evolutionary biologist to know better. You do not have to be Richard Dawkins to appreciate the fact that the God of War is the God of the Old Testament. The spiritualisation of God was the first revision. The spiritualisation of Jesus Christ was the second. It just was not helping any Protestant to be told Jesus Christ did not die on the cross. It was philosophy gone too far. All philosophy is a lie, and the greatest philosophy is the greatest lie. Honest Iago was not there to pity or save the Moor. He was there to condemn him. He was in the theatre to save London. Meanwhile, the Moor consoled himself that murder and honour amounted to the same thing.

The indignant Iago is an avenging angel. He simply executes judgement, "Demand me nothing. What you know you know". The villain of the play is Othello and yet he called Iago a demi-devil. But everyone in London knew who the real devil was. His downfall was good for every existential, every historian, and every Protestant. It was good for progress which the philosopher was in the habit of dismissing because of his second-hand philosophy. The Moor was convinced he was a spirit being and, in his operating system, the just lived by faith and honour. His sense of honour drove him to murder. He made honour a thing of tyranny. Iago turned his honour against him. No good thing can come out of philosophy. No one really wanted to know what was going on inside the mind of an animal. Suffice to say the system was unnatural and he had trapped himself in philosophy.

Iago's personal grievance does not justify the judgement visited on the Moor. His revenge is severe and disproportionate when considering the many innocent lives destroyed to save Venice. But there was a Moor on stage and the undercurrent of hate boiled over. Othello the lying philosopher was reason enough for Iago to act as Christ's avenger. If not Christ, then it was the universe he was avenging. Philosophers make a trophy of the universe for their creator-God. The arrogant Moor knew absolutely nothing about sub-atomic particles and quantum mechanical devices. He had no idea of the increasing and incredible distances between galaxies. Yet, he was willing to bet his soul on an omnipresent philosophical entity which he was sure had emanated the universe into existence. He was not willing to wait for science. He had revelation and gnosis.

Iago's hate and contempt for the philosopher won him the sympathy of the audience. Total hate was what they expected to see. Christ expressed hate for Romans because they dared to occupy the holy land. In Galilee he expressed hate for Pharisees. In Jerusalem it was hate for the Sadducees. Christ had nothing but contempt for Hellenistic Judaism, a philosophy. Paul the non-apostle hated *Christ-after-the-flesh*. He passed himself off as an apostle and invented Hellenistic-Christianity. He despised the Galilean apostles. Paul hated Greek apostles because Apollos who was a real apostle was gay. Corinthian Christians hated Jewish Christians and refused to give assistance to Jerusalem at a time of famine. Iago hated all philosophy because it is all lies. Jesus is not bigger than the Milky Way and even Mars is bigger than him. Christ said he was God. He did not say he was omnipresent.

The Reformation was an attack on philosophy. The Reformation came out of the gospels. The gospels have an existential Jesus. The gospels are based on a historical Christ. This was a shocking discovery for everyone new to books and learning to read. God could not be more humanistic than God in Jesus

Christ. It was convenient that the Moor was a philosopher who was neither Catholic nor Protestant. It made him the simplest, purest, and best representation of philosophy. The brand did not matter. The Moor made it easy to target philosophy. It was not possible to hate the Christian. Hate was the only Christian way of dealing with the Moor and that meant disassociating as completely as possible. Christ was not interested in governing the world. He was for his friends the apostles, a closed group. Protestants and Catholics were left to deal with each other and sort out a working relationship. Queen Elizabeth gave them an England to grow in.

The despicable Iago became the hero of the play the moment he confessed to hating the Moor. The play has villainy at two levels. His punishment for the philosopher was exquisite. The audience had to follow the plot right to the end to see Othello humiliated. The Moor who fooled Venice was in the end hated and rejected by the Venetians. Othello was not Russian, but he was someone set on deceiving himself. He was on the constant lookout for ways to feed his delusion. He was a slave to his philosophy, the source of his faith and honour. Venice was a better place without him. Iago the defender of the Reformation not just hates the prophet but takes action to expose and crush him. He takes the audience with him on the journey of retribution. The Moor chokes on his honour. But he is unrepentant to the end and blames Iago for his immoral philosophy.

Ancient Greece gave a good account of supernatural dualism. They believed knowledge of invisible things which explained visible things made someone a prophet. The universe was intelligible and that led to the natural sciences and prophets. Meanwhile, Romans crucified people for sedition. Find the sedition to find the historical Christ. Its omission is nothing short of perverting the course of justice. Omission is an invaluable tool for philosophers. Before omission there was a legal statement and something lawful to say about Christ. Criminality was written out of the Jesus story. Supernatural dualism declared, "...henceforth know we no man after the flesh: yea, though we have known Christ after the flesh, yet now henceforth know we him no more". Christ-after-the-flesh was omitted, and Christianity became total philosophy. Philosophers did what they needed to do to assimilate Christ into their cosmos, and civilisation.

Ignatius of Antioch, a First Century church leader might have been talking about St. Paul when he said, *If, therefore, those who were brought up in the ancient order of things have come to the possession of a new hope, no longer observing the Sabbath, but living in the observance of the Lord's Day, on which also our life has sprung up again by Him and by His death-whom some deny.* Yes, there were some like me who denied the Lord's death. The most prominent Christian ever to deny Christ's death was Apollos who was not just a church leader but also one of the chief apostles. He was a friend of and personally deputised by Christ. Apollos could boast a personal relationship with God through Jesus Christ. He was a man taught by God. Apollos denied the death of God because Christ boasted he was invincible.

Death was for show. Invincibility was for show. It was to provide vindication for his friends. The legal contest and conclusion led to the legal statement which adorned his cross. It identified his crime and his sedition. The cross testified who he was, and it came from Rome a third a party to the quarrel between Israel and the God of Israel. Roman law was just as much on trial, and the legal precedence was of interest to every Roman citizen. Especially, those who bought Roman Citizenship for protection and liberties. The apostles had a front-page story which spread word of mouth and only they could explain from point of view of the God of Israel. St. Ignatius was yet another philosopher who thought Christ belonged to philosophy because divinity came out of philosophy. His intellectual discipline forbade him to think blasphemous thoughts. The idea of God-after-the-flesh was the equivalent of a spiritual brain aneurism.

Philosophers needed resurrection to prove spirituality the operating system. They needed the death of Christ for resurrection's sake. Ignatius of Antioch failed to mention the other school of thought. A few centuries later came the Moor. The Moor looked for answers to get him past the shape-shifter Jesus who was bigger than the solar system. The Moor was like St Ignatius and St Paul in that he too had no idea what Jesus of Nazareth was about and like all other philosophers before him he really didn't care. Philosophers were all about making progress toward a fairer and more inclusive society. Romans were materialistic and known for their overindulgences. Meanwhile, philosophy provided society with a social conscience. Meanwhile, the best we got from Christ was moral advice, do unto others as you would have them do unto you.

Christian theology does not come from Christ because he was not a philosopher. The historical Christ was of no use to philosophers who wanted to engineer a new social reality and a better future for themselves and their supporters. It was increased consciousness of a kind. It attracted the curious and the hopeful. Paul and Ignatius filled in the gaps in theology. They created a huge gap by identifying and leaving out Christ-after-the-flesh. They filled the gap with spiritual mechanisms. An omnipresent Jesus is hard to justify. Christianity set the problem and the Moor attempted to solve it. The Moor was about an improved civilisation for Moors, and he too looked to socio-spirituality. The Moor was born long before the Reformation when there was only Catholicism. He went his own way into philosophy because he did not want a Christ with a criminal conviction to his name. He looked for another narrative.

The Moor found what he was looking for courtesy of St. Ignatius. He picked up on the comment, *His death-whom some deny.* He investigated the possibility that Christ did not die on the cross. The new philosophy was applied retrospectively, it was projected into the past. The past was rewritten. It was revision. Although St. Ignatius was not directly addressing the issue of a super-sized Jesus it was relevant. When Ignatius said Christians denied the death of Christ it was obvious to the Moor that the righteous Christ was saved from the ignominy of the cross. The Moor had two issues with the Catholic Christ. The Moor was correct to reject the super-sized Jesus and associated faith, philosophy, cosmology, and civilisation. The multi-galactic and inter-galactic Christ was gone. The Moor like the original philosophers of Ancient Greece lived in complete ignorance of the Israelite God of War. Revision went back beyond Jesus and all the way to Abraham. Abraham was branded with his philosophy.

Shakespeare assigned to the duplicitous Iago the phrase "I am not what I am". If this is derived from the Bible's "I am that I am" it is hugely significant. If circumstances were different Iago would have said, "I am that I am". Exactly what is the Bible expert trying to say by using this famous line. Was Shakespeare a reliable intellectual on a par with Richard Dawkins and Stephen Hawking? I am an existentialist and I see an existentialist red flag. Shakespeare was not an evolutionary biologist and he lived long before discovery of Big Bang and the separation of the galaxies. If "I am that I am" and "I am not what I am" are two sides of the same coin, then it was not Richard Dawkins who first discovered the God of War in 2006. It was William Shakespeare who flagged the God of War in 1623. He managed to get it past the philosophers.

History revealed the God of War and the God of War revealed history to be the preferred method of revelation. The same went for Christ who relied on witnesses and testimonies to spread the news in one form or another. Over one hundred thousand Passover visitors returned home with a story to tell. Some spoke of messiah, and some spoke of Christ. Within days the theocracy, the Temple, the official visit, and Roman law became topical. For many in the Greek world it was the Unknown God making himself known. Christ was an international curiosity. Roman law had declared him a deity. Only Christ's apostles could offer some kind of explanation. They knew the how and why. The Greek Synagogue in Jerusalem played a key part in events as did the Sanhedrin and the Roman magistrate whose judgement on the matter made headlines.

Philosophers were not happy about losing ground to the God of War. They fought back. They employed St. Paul in special capacity by making him the Chief Persecutor of the Church. They twisted the narrative to make it about betrayal. They made it about a failed military coup. There was a political stand-off involving messiah and the Roman Governor. Those who knew what he was about decided not to air the issues in his absence. It was crazy to think he was God Almighty. Christ-after-the-flesh was an aberration. They did not want to assist his cause. They omitted him from memory and started arresting his followers for blasphemy. It was already blasphemy to speak the word and name Yahweh, so Jerusalem was able to absorb the visit of Yahweh without too much inconvenience. Blasphemy meant no one could talk about it.

It is one thing to revise history in a bid to generate faith. It is something else to destroy history completely. Having said that, none of the crimes against humanity committed by wicked Gnostics matter to us if the God of War does not exist. The crimes by Hellenistic Jews, Hellenistic Christians, and Hellenistic Moors seem to be against each other. The Temple Mount was the place on Earth where the God of War declared himself to the world. It is also the place where he made it clear to Pilate that he posed no threat to Rome. The kingdom of the God of War was not of this world. The magistrate needed to charge him with sedition, but the accused was not making it easy for him. The world was none the wiser about the God of War till Richard Dawkins found him in 2006. Even then we failed to understand what the God of War meant till 2010 when Stephen Hawking became aware of him.

Christ was aware of the God of War because he was the God of War. By the time the Moor discovered philosophy the God of War was already a forgotten victim of Aristotle's. The Moor did not have opportunity to reject the God of War because he had no knowledge of his existence. Existential was not in his vocabulary. He was a philosopher. Philosophers did not want to see a God of War. They wanted to see God in principle, the agent responsible for emanating the universe into existence aka the Creator. Jesus Christ was a revelation of history but that could not be said of the God of War till 2006. When Stephen Hawking said the God of War is the God of the Old Testament it was thinking without philosophy. Shakespeare was equally good at thinking without philosophy. The Moor could not do that.

The Moor has always been famous for one thing. He grabbed the Temple Mount for himself but by doing so he made himself the enemy of all Jews, Christians, and existentialists. Military dominance meant he could get away with the theft of their history. King Herod built the Temple Mount and gave it to the Jews. The Moor took it by force and as far as he was concerned his new acquisition was worth all the hate. All Jews, all Christians, and all existentialists will always hate the Moor for what he did. There was protest and in his arrogance the Moor said to them, "Demand me nothing. What you know, you know." In other words, deal with it. The Moor chose the same auspicious spot as the God of War to announce his arrival to the world. The Moor made his occupation of the Holy Land a permanent one by planting a religious icon on the Temple Mount. It was a declaration to forever remind Jews and Christians of the power of the sword.

The Moor appropriated the Temple Mount and like the good philosopher he was he ended up destroying all trace of the God of War. He ended up consolidating philosophy and rededicating the Temple Mount to the non-existent philosophical entity known to millions as the creator. It was philosophy with no scope of a God or War. The magic of prophecy meant he could apply his philosophy retrospectively and he did. Philosophy again hijacked the past which used to belong to gods of war. The creation of the universe placed the creator ahead in time to all other gods. For anyone who recognises the God of War to be an existential creature there is nothing fouler than the philosopher and creationism. For anyone who in any vague way recognises the existential Christ there is nothing fouler than the philosopher and creationism.

Finally, there was revenge. Iago had his pleasure. Iago believed it was OK to do unto others as they do unto you. A deception was played out on the Moor. It left the Moor drowning in his perverse sense of honour. The sweetest words Iago ever spoke were "Demand me nothing. What you know, you know." It was too late to repent because it was judgement day for Othello the Moor. Iago said, "I hate the Moor." It was all about the hate the Moor set off when he said to existentialists "Demand me nothing. What you know, you know." Hate caught up with the Moor. Iago was willing to suffer torture, death, and Purgatory to show the Moor just how much he hated him. Hate belongs on the moral compass and a moral compass is one thing philosophers do not have. They are hypocrites. When they hate the flesh, it is self-loathing. When spirituals hate the world and fill it with spiritual mechanisms, it is social engineering. When they hate the devil, it is delusion.

The First Folio of William Shakespeare is an existentialist work and therefore the first serious alternative to philosophy and revelation. The First Folio is the foundation for a future civilisation, a brave new world.

# Chapter 5

# Philosophy

I was deceived along with many others. Billions have been tricked into believing an untruth. It is all of us affected by the falsehood which came from Anaximander of Miletus (610 BC - 545 BC). The great philosopher came up with the *primary substance,* more commonly known as the Creator. Thus, making him the world's very first theologian. He did not mean it as a lie, but his mistake is a falsehood we must deal with. Anaximander said, *"All things arise from a primary substance and return to it, unless they pay fine and penalty for their iniquity".* Creator seems to be the must have qualification for God to be God. Anaximander's primary substance found its way into the Old Testament. Hence, Hellenistic Judaism.

The mistake went undetected till 2006. The unsuspecting Richard Dawkins read the Old Testament only to discover Yahweh the God of War. The Bible has 2 concepts of God in Yahweh the Creator God and Yahweh the God of War. I can imagine the God of War reprimanding Anaximander, *Ye do err, not knowing the scriptures, nor the power of God.* But in truth, his being Greek meant the academic was of no interest to the geo-political tribal God of Israel. Dawkins discovered the primary substance to be a thing of terror because it gave rise to primary substance practitioners aka the faith family. The faithful had pushed the God of War into the margins. Dawkins knew he was not imagining the God of War but there was so much denial coming from the faith family he needed reassuring. Stephen Hawking (1942-2018) vouched for his comprehension skills.

The God of War acquired two formidable champions in Richard Dawkins and Stephen Hawking. It was because science managed to find the other primary substance. The hot dense primordial soup of matter found at the base of Big Bang is the real primary substance. Particles are the building blocks of the universe. The faithful gave a nod to Big Bang but carried on relying on the belief that an omnipresent creator had emanated an omnipresent universe into existence. They were not letting go of the Creator. Over a period of four years Dawkins began to realise the Bible's God of War is one of the most effective weapons militant atheists have in their war against the formidable faith family.

Anaximander's primary substance evolved. Divinity or theology is an intellectual discipline. The most famous contributors being Socrates, Plato, and Aristotle. The error was compounded. Theist and theologian Aristotle (c. 384 B.C. to 322 B.C.) wrote bestselling books about metaphysics which is how come we know so much about the history of the primary substance. But there is another source. The Old Testament is a compilation of 39 books. It was obviously compiled after the 39 books were written. The Old Testament was compiled in the time between Malachi and the gospel of Mark. The window is 400 years. Hellenistic Judaism was the editorial policy applied to it. Hellenistic meaning derived from Greek philosophy.

Aristotle's, Plato's, and Anaximander's books amounted to the Word about God. The chief mechanism for finding out about God was the natural sciences. Science or the idea of an intelligible natural world was good, but the idea of a scientist creator was silly. Intelligent Design wrongly assumed there was an intelligent designer. Aristotle was a prolific writer with international bestsellers under his belt. He is arguably the most influential Greek that ever lived. He worked with Alexander the Great in spreading Greek culture around the world. Back in the day "Hellenistic" was an international phenomenon and hard to ignore. In 167 BC Hellenism was forced on the Jews. In due course the Jews had their own Word About God, but they called it the Word of God and contributors were known as the prophets.

The natural sciences were a popular subject for study. It was popular in many parts of the Greek Empire. The universe was intelligible but since so many were unlearned there was for them the method of revelation. Revelation was assigned to the prophets of Yahweh. It was assigned to Moses. Anaximander gave the world natural theology. The prophets made it into revelation theology. They all agreed creation had a creator in principle. Aristotle was the Einstein of his day. People the world over could do the same experiments and make the same scientific discoveries. Universal truth was universally appreciated. But it was flawed. The universe's operating system was thought to be spirituality. Their natural world was for them evidence of the primary substance.

Primary substance practitioners created the primary substance to go with their science and their science justified belief in the primary substance. Divine philosophy emerged and was made into a working system. Man, who was alienated from the Creator could be reconciled with gnosis or knowledge of God. Aristotle like other philosophers had his own version of the primary substance. The Holy Trinity is a more recent version. Richard Dawkins another leading scientist wanted to write yet another book and for that book he needed to research the Old Testament. He read it. But to his surprise, he could not see Yahweh the Creator God. He was blinded by Yahweh the God of War. His intellectual discipline was not theology, he was without faith. Meanwhile, faith had evolved into theocracy.

The biologist had no problem seeing the biological God of the Old Testament. But it was too silly an idea to merit serious consideration. His book was going to be about the God delusion and the God operating system. His focus was on mechanisms and methods underpinning the delusion. The militant atheist was naturally lacking in faith. He had his own idea of how the universe had come into being. His operating system was Big Bang. He relied on sub-atomic particles being the evidence of things not seen. Most of the stuff about the Primary Substance and Creator God propaganda is found in a few verses of the Old Testament. Dawkins latched onto history for an understanding of God. He ended up with an angry and violent ethnic cleanser.

Dawkins was dismissive of God the psychotic delinquent because of common sense. Common sense is the greatest enemy the spaceman deity has. FYI the Big Bang theory was not available to either Anaximander or Aristotle. They did not have the technology. They had the natural world and their own common sense. Stephen Hawking and co-author Leonard Mlodinow in *The Grand Design* acknowledged the fact that cosmology and theology share the same source. We know the origins because of Aristotle. He was a top theologian, a top scientist, and a bit of an historian. He was the first person to distinguish between physics and metaphysics. His books put the Hellenistic in Hellenistic Judaism. Metaphysics was given its own history which went all the way back to Anaximander, the man who not so much found as invented the primary substance.

Although Anaximander invented the primary substance the people of Miletus already believed spirit was a substance. They believed in reincarnation and that man is spirit. They believed in faith in good works. The primary substance was a sort of super spirit. It made sense of reincarnation and faith in good works. There was someone in charge. Two thousand and six hundred years ago spirit was part of the natural world. It was possible to look into the mind of God. We look back and see supernatural dualism. They saw it as a supernatural singularity. They got it wrong. Aristotle who was more often right than not contributed to theology. Theology went unchallenged till 1939 when it ceased to matter. With war came year zero. After the war theologians did not know any better than to carry on being theologians.

Science is universal truth. A science fact is true anywhere in the universe. It is a fact regardless of a person's colour, religion, and time. Theology was once thought to be universal truth. Philosophy was flawed from the very beginning. A system was built to deceive itself. Divine philosophy was the mistake of one man. Anaximander started a scientific enquiry to satisfy scientific curiosity. He was prompted by Thales of Miletus (c. 620 BC - c. 545 BC) who proposed everything is made from water. It was a time when there was nothing Hellenistic about Judaism or Christianity or even Russia. In their bid to understand the natural world elements such as water, earth, air, and fire were proposed as the building blocks of the universe. The four elements failed to explain man, reincarnation, and the impulse to do good and right.

The competition was on to be the first school of philosophy to come up with the theory of everything. The theory had to incorporate re-incarnation and transmigratory spirits which were popular beliefs of the day. It had to explain the life-death-life cycle. They did not distinguish between natural and supernatural or physics and metaphysics. Hawking and Mlodinow of the 21$^{st}$ Century traced cosmology back to Anaximander. They included the famous quote, *All things arise from a primary substance and return to it, unless they pay fine and penalty for their iniquity.* The father of cosmology is at the same time the father of theology. When Aristotle made the distinction between physics and metaphysics, he did so without challenging the existence of a primary substance. The cosmological argument for the existence of the primary substance was not challenged by Aristotle.

A purely physical universe was not arrived at till the 20$^{th}$ Century. Matter was the mysterious element, and it could be traced back to a hot dense primordial soup. Hawking was one of the fathers of the Big Bang theory. In 2010 he interpreted the laws of science to make it a lawful thing to say that God did not create the universe. In 2006 an evolutionary biologist used literacy to make it lawful to say God is a spaceman. History was once again the primary intellectual tool for understanding Yahweh the God of War and ethnic cleanser. In Russia nothing could be heard over the sound of propaganda. Worse still, 2013 saw the Kremlin pass laws making it punishable to insult people's philosophical feelings. The Russian people swapped freedom of speech for freedom of faith. It was like Judea in Jesus's time all over again. It was then unlawful to say Yahweh.

Both Aristotle and Hawking acknowledged Anaximander's great contribution to the natural sciences. But there was a downside. He ended up leading the whole world astray with his religious ideas. Aristotle was first to sense grave danger. We of the 21$^{st}$ Century know from experience that theocracies are totalitarian in nature. The Russian Federation led by the Russian Orthodox Church in 2022 threatened the world with nuclear holocaust. Romans were first to realize Christianity is a philosophy. The historian Tacitus said about them, "all of whom were condemned, not so much for the crime of burning the city, as their hatred of mankind." Not all Christianity was philosophy and therefore not all Christianity was fascist but then the Roman Empire was a big place. Russians are led by faith and propaganda and the reason is Christianity. Stalin was a theologian, and he came up with the patriotic war which was about as useful as a war with Beelzebub.

Socratic philosophy was a fascist threat, and this was obvious to Aristotle. Academic progress and freedom were Aristotle's idea of philosophy. He identified what was wrong. He came up with the "Unmoved Mover" and the "biological soul" to stop fascism in its tracks. But the system which was bent on deceiving itself set out to deceive the world with renewed vigour. A philosopher strived to attune his mind with the primary substance. Emperors became "God manifest." Faith could do so much and all because of Socrates whose faith made God a part of daily life. Theologians of today want God to be a part of daily life. They ban the teaching of evolution and science which suggests they might be wrong.

Faith is all pervasive in Russia, Propaganda is all pervasive in Russia. Lies and deception are all pervasive in Russia because the name of the game is manipulation and social engineering. Ivan the Terrible built a beautiful cathedral in Moscow around the same time the beautiful Elizabeth came to the throne in England. Rulers have faith in God and sometimes they have faith in mutual benefit. It is a mixed bag. Socratic philosophy was destroying his world and Aristotle hit out as hard as he could. Aristotle was first to say the primary substance did not create the universe. He was first to say heaven and earth are incidental and not a direct creation. He introduced the biological soul to get rid of the idea that we are spirits bound to return to the divine source. Greater good was done by Aristotle through anger than through being a purist.

Religious people still find it hard to deal with Aristotle's idea of a God exists, but is not the creator, and he is not our creator. The attribute of creator was removed from the primary substance. The Unmoved Mover suggested creation was incidental and even accidental. This is not far removed from the Big Bang principle that the universe is a spontaneous creation.

It is true, God did not create the universe. He is not our Father. The only person who can call him father is the God of War. The only people who can call him father are the people of the God of War. A few Galileans managed to force Christ into praying. It was a communal prayer. It was Christ's prayer, and his God became their God, and his Father became their Father.

Aristotle thought Socratic philosophy was rubbish and the ruin of humanity. It was the pursuit of delusion. Socrates had followers who wanted to know the meaning of raw existence. They were after all spirits in the material world. They wanted to know if it was possible to lose all sense of the material. Sounds terrifying! The pursuit of science fell by the wayside. Philosophy needed to be made safe. His "biological soul" helped distinguish between natural and supernatural and created a dividing line. Socrates claimed God had a part to play in daily life. It was a big part and even a togetherness, a sort of mind mild or spirit meld. Socrates took the primary substance out of the classroom. Aristotle did his best to have it returned to the classroom. He could not believe Socrates had people praying to the primary substance.

Philosophers who believe God plays a part in daily life are threatening the end of the world. The latest Church diktat is the only one that counts. No one is alarmed by the contents of the diktat because moral responsibility is transferred to the head of the church. Church has control. He who controls the Orthodox Church attracts the Kremlin. The system and methodology of church has led to Kremlin Christianity. The just shall live by faith and that includes nuclear holocaust. Thankfully, not all conscripts were able to commit murder, not even for the Patriarch or the President. Spirituality did its job which was to create an existentialist crisis and a test of faith. Many discovered consciences and the impulse to do good and right kicked in.

Haters of the faith family acquired a friend and champion in Stephen Hawking who said, *Spontaneous creation is the reason there is something rather than nothing, why the universe exists, why we exist. It is not necessary to invoke God to light the blue touch paper and set the universe going.* The primary substance was a mistake unless you mean the hot dense primordial soup of matter. God is not a variation of the primary substance. God is not the hot dense primordial soup of matter. If God does exist, then we must look beyond the primary substance and the hot dense primordial soup of matter. Although there is good reason to believe Catherine Parr, Elizabeth Tudor and Willian Shakespeare believed in the historical revelation of God in Christ it opened too many gaps in philosophy. Philosophy came with cosmos, the natural sciences, intelligent design, civilisation, and divinity.

Aristotle was not interested in gods of war and skygods. When he said God did not create the universe and it was not possible to have a spiritual relationship with God, it left him with an omnipresent philosophical entity short of two attributes. It left primary substance practitioners with choice of degree of spirituality. Aristotle made philosophy complex, exciting, and engaging. The biological spirit divided philosophers into two groups, the Socratic and the Aristotelean. The professor of biology understood biology well enough to say man is wholly biological. He needed to go a bit further and ask himself how it was possible for a biological spirit or soul to come into existence without God's involvement. His research would have enabled him to say it was like other organs and a result of embryonic development. Another way of saying something is developing is to say it is evolving.

For Aristotle philosophy was about science and education. It was about social betterment and progress. It was about improving the mind and one's quality of life. He offered his students a university education and career prospects. Meanwhile,

Socrates wanted everyone to be a theologian and enter a spiritual relationship with God. Socrates made God into a personal God. Aristotle acknowledged the existence of the primary substance and considered it a human privilege to know that a creator exists. But for him it stopped at Intelligent Design. Sadducees were Aristotelean because they did not believe in resurrection. Pharisees were Socratic because they did. Sadducees rejected the notion of supernatural beings like angels and a supernatural messiah. Not all philosophers believed a spiritual relationship with God was possible.

Socrates the preacher and Aristotle the educator was about social engineering but of different kinds. If we go back to the first theologian, we find a primitive idea of God. Anaximander said, *All things arise from a primary substance and return to it, unless they pay fine and penalty for their iniquity.* The agent of creation did not have a personality to speak of. They started with a process which applied to spirit beings. Reincarnation applied to spirit beings. They believed in dualism, a supernatural dualism. They did not have Richard Dawkins who acknowledged dualism but had his own way of describing the phenomenon. He said, "*Our innate dualism prepares us to believe in a soul which inhabits the body rather than being integrally part of the body.*" The scientist is brilliantly succinct when describing exactly what Russian Orthodox Christians make of it. "*The dualist believes the mind is some kind of disembodied spirit that inhabits the body and therefore conceivably could leave the body and exist somewhere else*". Innate dualism is part of the human condition. We all experience it.

Man is not spirit and therefore there is no spiritual warfare going on. Spirit and flesh are not at war with each other. That is not a correct description of what is going on although the church of St. Paul does its very best to make it appear so. Socrates invented spirituality around his dualist experiences. The public speaker could hear himself think while delivering speeches to a live audience. He planned and rehearsed what to say but sometimes thought interrupted and redirected him mid-flow. It was like someone else was in control. Socrates used to talk to himself and one day while talking to himself he realised he was talking to his other self. But he interpreted it as supernatural dualism, and he thought it was the primary substance giving him guidance. Innate dualism, biological soul, disembodied spirit, and double entity are some of the descriptions given to the same phenomenon.

Socrates made it his business to prepare his spirit for afterlife but his was not faith in good works. His was not faith in process. His was a faith in faith. Plato and other students had faith in the faith of Socrates. The philosopher involved the life-giving Primary Substance in every aspect of daily life. He made a parent out of the primary substance. Parent made a personality God of the primary substance. Aristotle the scientist thought intelligent design was plenty philosophy to be getting on with. Science was to liberate man by giving him a fuller understanding of himself and his world. It was the reason for education, for the schools of philosophy. The natural sciences validated the Primary Substance and acted as evidence for the same.

We got to the end of philosophy in 2010. There was a new model of the universe, and it was called the Big Bang theory. Finally, the God of War of the Old Testament and Ancient Israel came into his own. History revealed an unpleasant ethnic cleanser who did not like Gentiles and that included Richard Dawkins. What little was known about the violent bully provided no traction for a sustained debate. According to the Bible itself the God of Israel started life by killing ancient Egyptians and Ancient Israelites. There was nothing nice to say about him. Dawkins blasted away the treachery of philosophers to reveal the God of Embarrassment. The militant atheist had not converted or anything like that. His own idea of God was that of a Flying Spaghetti Monster, absurd, irrelevant, and impossible.

Dawkins like Shakespeare came across a historical revelation of God. The difference being that he was not trying to understand the historical revelation in context of the philosophical revelation. At thirteen, Dawkins took a leap of faith. In 2006 he took a leap out of faith. If the creator God does not exist, then faith is no more the *evidence of things not seen.* Faith is evidence of delusion. Spontaneous creation replaced divine creation. Physics replaced metaphysics. Sub-atomic particles replaced the theory of emanation. Shakespeare lived in sensitive times, and he could not say the primary substance is a mistake. The Reformation was not a licence for introducing the God of War. Anyone believing in historical revelation had to keep it to themselves. It was philosophy which was invincible till 2006.

Something strange happened to Amanda Holden the TV star in 2011. She experienced out of the body except it was not that. It was more than that. She experienced having two bodies at the same time. One Amanda two bodies is the best I can describe it. Giving rise to the possibility that Dawkins might have got it wrong. He said, "*innate dualism prepares us to believe in a soul which inhabits the body rather than being integrally part of the body.*" But according to Holden's experience of dualism *a soul inhabits the body and is integrally part of the body.*"

# Chapter 6

# Historical Revelation

*Many have taken in hand to set forth in order a declaration of those things which are most surely believed among us, even as they delivered them unto us, which from the beginning were eyewitnesses.*

From the gospel of Luke

We do not have any of the original transcripts Luke had access to. Not all Galileans could read and write but they made sure there was a declaration telling their story. The documents were delivered to the designated church leaders. None of the contributors were historians or journalists. Luke had to allow for myth and superstition. He had to compensate for the gaps in philosophy and theology. Messiah had his own ideas, and he was a bit of an anomaly. Making a book out of all the various accounts demanded selection of the material to be used and some sort of editorial policy. We do not have any of the original transcripts, but we do have the next best thing. We have Luke's gospel.

Stories could be cross referenced and verified. The real challenge was to get everything in correct order. The eyewitnesses did their best *to set forth in order* the things they saw and experienced. Luke needed order for the story. His aim was to write a book. As you might expect, no one had the full story. Not even the twelve disciples who obviously had a large part of the story. They were the main contributors. The twelve had their own inner circle and therefore Peter, James and John had the most to tell. Luke even had a declaration from a Roman Centurion. Simon the Pharisee was happy to contribute. There were miracle stories like that of blind Bartimaeus. The women who played a big part in ministering to the Lord posted a group declaration.

The biggest surprise was the contribution from the wife of Pilate. Stephen the First Christian Martyr made a submission. The story of the Jews was supplemented with the story of the Gentiles. Luke had a lot of information to get through. Lucky for everyone he was the right man for the job. He was a professional and could make use of his intellectual discipline. The doctor knew the difference between cause and effect. There was a good reason for Luke to write his gospel when he did. Theophilus was thinking about converting but all he had was random bits and pieces on Jesus Christ. Luke needed to do a quick write up. He told the Roman officer, *It seemed good to me also, having had perfect understanding of all things from the very first, to write unto thee in order.*

Luke based his gospel on eyewitness testimony because he was not there in person when Jesus of Nazareth campaigned in Galilee. The stories he chose and the order he gave them were for the benefit of Theophilus. He could direct the narrative to convey important events and beliefs. He did his best not to make things more complicated than they already were. Theophilus understood something of the Jewish tradition of messiah. Luke provided information to help the Roman conclude messiah was friendly toward Rome. Pilate was personally responsible for involving Rome and fault was not with the office of Governor. Luke skipped over Mary Magdalene to make messiah more of a traditional character and less of a social reformer. It was not going to help if Rome was to feel threatened by the church.

Luke had a purpose for his gospel. *That thou might know the certainty of those things, wherein thou hast been instructed.* All things in the eyewitness accounts leading to uncertainty were omitted. Luke did not search out Pilate or Caiaphas for comment. Luke had the advantage of working with St. Paul. Missionary work identified some of the questions which needed to be answered. Luke the editor gave importance to theology over all the anecdotal evidence. The gospel needed to be edited and have some kind of narrative. Not everyone was going to be happy with the final draft. Writing a gospel was always going to be controversial. In the end the remarkable thing was not perfect understanding but just understanding. It was a huge achievement to make a book out of the declarations.

St. Paul was another with an overview. He too was not an eyewitness, but he came across many of the eyewitnesses and he was a library of information. You could say Paul had seen and heard more eyewitness declarations than anyone else. It was his job as Chief Persecutor of the Church. He lived with Peter and James for fifteen days. He was a worker in Antioch where Jews and Gentiles began working out a new relationship. It was inevitable for the many eyewitness declarations to be compacted into a gospel. I have written monologues in a bid to understand what it was like before the first gospel was written.

*Joseph of Arimathea*

I am in Bethlehem today because of a four-hundred-year-old promise. The Scriptures ended with this exciting promise, *the Lord, whom you seek, shall suddenly come to his temple.* There is an even older promise. *Bethlehem... out of thee shall he come forth unto me that is to be ruler in Israel.* John the Baptist is the voice in the wilderness calling on God's people to prepare for the visit of the Lord. The forerunner of the coming Lord fulfilled the first sign that the great and terrible day of the Lord is upon us. I left Jerusalem for Bethlehem on his pronouncement that Messiah is about to be revealed. Some rare event ordained by God is about to happen and it started here in Bethlehem.

John the Baptist is following the lead given by Moses who prepared his generation for a divine revelation. They were cleansed. The prophet of the wilderness is washing the people of his generation in readiness for another major reveal. John said the one to come is God's personal choice. God has once again gone through the nation in a bid to find a second David. The Baptist succeeded in that Jerusalem is buzzing with talk of messiah. He is a hidden figure and till he is outed he remains hidden maybe even to himself. The response to the Baptist has been twofold. There is earnest debate and a revisiting of messianic prophecies. But there is also theology which seeks to clamp down on idolatry and belief in supernatural beings.

It is exciting to think that oils which fell on David one thousand years ago also fell from him onto the ground, possibly right here where I am standing. Bethlehem too is anointed and the subject of divine intervention. I might not live long enough to see the new king but being here I feel a connection with him. It was Jerusalem's Greek synagogue which insisted on exhaustive debate to resolve issues they were having. Some say messiah is divine and not simply divinely appointed. Some say there is no such thing as prophecy and messiah is born of hope. Some it is a fable from a bygone era. The Greek synagogue was not satisfied with any of the answers. They insisted on keeping an open mind. They could wait a little longer and after which they will be able to make up their own minds.

Sometimes scripture fails to explain itself. Like when it says, *For unto us a child is born, unto us a son is given and the government shall be upon his shoulder: and his name shall be called Wonderful, Counsellor, the mighty God, the everlasting Father, the prince of peace.* Debate left us wondering how much could be believed before we started contradicting ourselves. The idea of a divine Messiah is readily dismissed because only God is divine and anything else is idolatry. We couldn't be sure if messianic scripture was meant to be taken literally. We went into it deeply, but nothing was resolved. Scripture kept its mystery. It is like we are not meant to know what is meant when it says, *Thy throne, O God, is forever. You have loved righteousness therefore God, even thy God, has anointed thee.* Sceptics warned of myths and fables not unique to Israel and to expect to be underwhelmed by the voice in the wilderness.

I must tell you about Apollos the Greek proselyte and one of the leaders of the Greek synagogue. He is the most intelligent person I have ever met. He is convinced writers of messianic scriptures believed in the reality of messiah. He says it is true there is a lot we do not understand but that is not a reason for entering into denial. Greek and Jewish theology is suspended by him so that the Baptist is not denied his opportunity to prove himself. Sometimes I think Apollos is desperate for prophecy to be true. He wants to talk face to face with God like Moses did. Some in the Greek synagogue are carried away with the idea of a divine messiah. They do not seem to mind that they are excluded by the Baptist and have no place in his nationalist movement.

Being a Pharisee I believe in angels. Being a Pharisee, I can believe in a supernatural messiah. But I cannot believe in a divine messiah. The nation must be careful not to make an idol out of messiah and bring the scriptures into disrepute. The prophets are guilty of hyperbole and overstating the importance of the office of messiah. Prophets, priests, and kings of Israel have often been bitter rivals. It does seem like the prophets have picked out one of their own for a future ruler. Not everything about messiah is unreasonable. His being a son of David is popular. There is no resolving the controversies and contradictions. There is no agreement on what can be expected.

One thing is for sure, the prophets have set a dangerous precedent. The Sadducees have their own way of dealing with prophecy. They represent the majority view and they do not believe in angels and skygods or a supernatural messiah. For them Messiah is an ordinary man, a descendent of David. The priests of Israel concluded the prophets of Israel meant to test us. Messianic prophecies are a test of Israel's proclivity for idolatry, an old problem. Put simply, it is nothing short of idolatry to make a man into a god. Pharisees and Sadducees agree that regardless of what our prophets have said this generation will not be relapsing into idolatry.

The Baptist enjoyed some success in Galilee. Unfair Roman taxes bred disenchantment and anger and his message was popular with some. It was fighting talk, but it was talk. The Baptist knew which verses helped vent rage. *In that day shall the Lord defend the inhabitants of Jerusalem; and he that is feeble among them at that day shall be as David; and the house of David shall be as God, as the angel of the Lord before them. And it shall come to pass in that day, that I will seek to destroy all the nations that come against Jerusalem.* The Baptist found verses which made his audience yearn for messiah. *Spare thy people, O Lord, and give not thy heritage to reproach, that the heathen should rule over them: wherefore should they say among the people, Where is their God?*

Jerusalem monitored the situation. Most people were curious and went into the wilderness to see and hear a real live prophet. Very quickly, everyone in Galilee knew about the coming of messiah and that was all it was about. In Jerusalem the only people distracted by him were Greek. The fact that they were proselytes counted for nothing. They were of no interest to the Baptist and his nationalist movement. I am like Apollos in that I too want it to be true. I want angels to exist. There was a lot of talk generated by the Baptist but nothing new was said. Academia has already exhausted the case for messiah. The idea of messiah is seriously flawed. Someone perfectly representing God needs to be exceptional and even if such a genius existed how would lesser mortals know his every command came from God?

Constant talk of messiah reminded people of all the failed messiahs. The thought of a Roman backlash and crucifixion on a grand scale is enough to subdue the urge for revolution. Messiah is an invitation to a bloodbath. Frankly, people are fed up with all the dangerous talk. They know the reality is that Rome is a superpower and after the Romans another superpower will take over. Monarchy is not the solution for the situation we find ourselves in. At least as a theocracy we can render unto God the things that are God's no matter whose empire we find ourselves in. As a theocracy we do not have to concern ourselves with political rivalries.

Speaking as a theologian I have to say the passages about Messiah are "inconclusive". We have no idea how to reconcile the contradictions and match prophecy with reality. But it is all a part of the Jewish legacy.

*Matthew*

The waiting is over. The Baptist made his choice. A minority paused to take breath. The majority who considered the wild man a curiosity, were happy for him. His critics, however, dismissed it another act of attention seeking sensationalism but this time leading to his downfall. The whole prophecy thing was messed up by the fact that his messiah came from Nazareth. Everyone knows the Bethlehem prophecy so what went wrong? It did not take long for the authorities to have the facts checked. Word reached Jerusalem confirming the much-anticipated reveal and that it was a Nazarene. For a while the whole nation was talking about him but in context of an elaborate scam. Scripture saved Galilee from being taken in by the Baptist and his messiah.

The prophets were obviously suffering from a breakdown in communications. It was back to business as usual for the people of Galilee. I am a tax collector and I know what it is like to be greeted with suspicion everywhere you go. The so called divinely appointed messiah was more famous for coming out of Nazareth than for coming out of prophecy. He was a non-starter. He simply could not be taken seriously. He was a joke. He was set up to fail. The nation with one voice said, "let's not waste any more time on him." Very soon the authorities knew all about him. They knew who his parents were and which synagogue they attended. For a while it seemed like the attention was too much for the new messiah and people thought he had gone into hiding. It was over before it started. The prophets were not all that.

Many were disillusioned and considered themselves fortunate not to have been taken in by the cruel hoax. They took a giant step back and wondered what it was all in aid of. It was not something he could hope to get away with. Then a month or so later word spread of his first public appearance. He was the only one in Galilee claiming the Baptist's mission a success. *The time is fulfilled, and the kingdom of God is at hand: repent ye and believe the gospel.* For a while he was the only one who believed the good news. The people allowed him to play out his fantasy. The Baptist was nowhere to be found. He carried on regardless of the prophecy fail. It seemed a thing so audacious, He carried on like it was Scripture that had got it wrong.

Then suddenly one day Galilee was ablaze with talk of a miracle. The chosen one was no more a social leper. Galileans were actively seeking him out because of the rumour of a miracle. They wanted to see for themselves. I wanted to see for myself. I set off as soon as commitments allowed. It was true and messiah had stepped out of the Baptist's shadow. It did not matter that he was a Nazarene. Jesus of Nazareth was winning recognition as a miracle worker. The prophets had their reputation restored. Suspicion gave way to popularity. Galilee was proud to have her own prophet. He always closed the miracle extravaganza with the haunting words, *blessed is he, whosoever shall not be offended in me.*

Our leaders had a thing or two to say about the prophet and from local synagogue podiums they provided guidance, correction, and warning. They advised caution and not to be taken in by miracles. His disqualification came from Scripture and there was nothing anyone could do about that. Scripture would catch up with him eventually and dismantle his platform of miracles. There was no question of calling him in and making him feel important. There was no point in indulging him. He was choosing to ignore prophecy and people did not care. It was going to be a matter of time but the decline in popularity was going to come. The authorities set about to increase the public understanding of prophecy.

Meanwhile, the public was enjoying listening to miracles. They were loud acclamations. God was playing to us. Life in Galilee got interesting. I was at work when the chosen one called on me. I already had a bit of a thick skin because of my job as a tax collector. I was not afraid of being shunned. I did not hesitate to join him in his revolution and the kingdom of God of the prophets. The prophet was not afraid of courting controversy. There was a miracle which stunned everyone because it came with the words *Son, thy sins be forgiven thee.* He knew he crossed the line. He excused himself with the boast *the Son of man hath power on earth to forgive sins.*

The Pharisees noted his defiance. They were doing their best to undermine him by telling him what he could and could not do. But the more they asserted their authority the more he asserted his, the unknown authority of messiah. The hope of working with Pharisees became a distant one. Without them on our side the only message reaching the people was that there was a dodgy prophet full of assumption. It was the same old story of prophets and priests vying for power. Pharisees command the loyalty of the people. They do a lot of good work. They are about national unity. They provide cohesion and messiah came over as divisive and sectarian. All attempts to sway them turned into angry exchanges. Some of the public started resenting us.

Pharisees carried on with their difficult task of national unity without messiah on board. They had the sympathy of the public who more and more were appreciating their quiet contribution to society. On top of everything else Messiah was acquiring the reputation of someone awkward and obstructive just for the sake of it. We were becoming more and more isolated. It was no use attacking their proud reputation because it only backfired on us. But he did and he attacked with gusto. We could have had social harmony. But the Son-of-man had his own priorities and he was not waiting for Pharisees or anyone else to catch up with him. His next move was yet more shocking.

We started campaigning on the Sabbath. It is the day of rest, and we were expected to cease from our usual labours which in our case was campaigning. It started with a healing and that drew criticism. Although the Baptist gave him an operator's licence, he was not stopping to ask the Pharisees for permission. He did not stop to pray or ask God for his permission. He was above all that. He could assume power and it was there. The Lord of the Sabbath was something new. Here was a new personage generated by the prophets. The personage of messiah became grander still. Yes, I did wonder why Jesus would not hold off campaigning on the Sabbath. We needed conciliatory gestures and not to provoke. I asked why he extended healing into the Sabbath.

I knew I could not anticipate the answer. I knew I was going to be surprised and I was. It was for the sake of an invisible audience. *Behold my servant, whom I have chosen; my beloved, in whom my soul is well pleased: I will put my spirit upon him, and he shall show judgement to the Gentiles... And in his name shall the Gentiles trust.* The only Gentiles I knew were the Romans who had invaded our country and for whom I had been collecting taxes. Gentiles have their own language, culture, and gods. Roman Gentiles are the enemy. Healing on the Sabbath was a method to increase visibility amongst Gentiles. They too needed to know he was Lord of the Sabbath. The healing of the man with the withered hand signalled to Gentiles their time had come. But there were no Gentiles present. Who was going to tell them?

*Simon the Pharisee*

I attended one of those open-air meetings. It was very confusing. I ended up knowing less than before. Thing is, if you don't know Jesus-speak you end up being excluded from I don't know what. We all know Rome will retaliate if there is a real challenge to her authority and supremacy. I assumed he was talking about Rome when he warned of impending persecution. *Blessed are they which are persecuted for righteousness' sake: for theirs is the kingdom of heaven. Blessed are ye, when men shall revile you, and persecute you, and shall say all manner of evil against you falsely, for my sake: rejoice, and be exceeding glad, for great is your reward in heaven.* Jesus has this habit of getting ahead of himself because he is not the threat that he likes to think he is.

We Pharisees have tried our best to understand what he thinks "messiah" means. It really is difficult because it is all based on assumptions. It suits everyone to have him spoken of in terms of a prophet but let's not forget he is yet to relinquish his claim on the throne of Israel. We Pharisees believe in the rule of law and no man is above the law and therefore no man is Lord of the Sabbath. We offer him no encouragement. You would think no true prophet would want to undermine the commandments. He is prone to fits of anger and we Pharisees know that only too well. Setting aside the Bethlehem fail there is nothing to indicate suitability for any kind of office. Miracles make neither king nor general. Cleansing the leper does not compensate for his lack of experience.

Those able to advise him on the law are conspicuous by absence. Any person benefitting from higher education poses a threat to the modest Jesus of Nazareth, not least the threat of embarrassment. We Pharisees are obliged to comment, and we can give back measure for measure. We are expected to be mindful and to contain any threat to peace. But in all this no one is being persecuted. The only thing persecuting him is scripture. The Nazarene has this incredible gift, but I am not sure he knows how best to use it. Of course, the people want to see miracles but very few are interested in a break away society. There are already too many sects. No one wants yet another social divide.

He seems to be doomed to majestic isolation because he insists on speaking as a king but not having earned any kingly stripes. The call to arms is simply not credible coming from a healer. Yes, with God on our side we can hope to destroy Rome the latest occupier. But do we really have to, and will it change anything? Eventually another empire will come and another after that. Rome seems to provide a convenient distraction for him because I learnt nothing about the alternative society he has planned. I have no idea what the Jesus monarchy will be like. Miracles draw men but his paranoia generated by his own foolish pursuit of the end-times repels them. End-times, apocalypse and Armageddon are not winning slogans.

He is a nationalist. He has no plans for what we do after Romans are expelled. The only policy I could detect was keeping them out. Society is changed and international relations is more important than ever before, it is a big part of the modern world. He is an extremist. His ideas are not practical. Nationalism is unworkable in the modern world. His demand that we take up the cross for his sake proves just how dangerous he is. Messiah is full of fight but so also is Rome. He can be subtle. Turning the other cheek is offering resistance and walking the extra mile is an act of defiance because it says his people fear him more than our oppressors. Any violent challenge is bound to bring the cross crashing down on us. He is someone who would risk the nation.

Any meaningful support for any rebel leader will simply hasten the inevitable backlash and that is why messiahs must be contained. It really will be too late once Rome decides to intervene. There is just too much to lose for the privilege of calling him Majesty. He is impossible. He gets angry with us for the disappointment he himself generates. Unrealistic expectations lead to stress and agitation. It is all too easy to blame our leaders. But he blames them for negotiating a peace deal. It is true the Temple functions within the imperial system, but our rulers have managed to negotiate important religious concessions. We are the world's first modern theocracy.

We have the freedoms necessary to render unto God the things that are God's. We have all we could want from the international community. He comes up with an alternative plan justifying it with, *Think not that I am come to destroy the law, or the prophets: I am not come to destroy, but to fulfil.* We felt obliged to ask a question or two. The laws of Moses are from God and so one assumes they are eternal. Eternal and change pose a problem and any transition needs to be overseen. Anyone not agreeing with Jesus is met with a practised rebuttal *you err, not knowing the scriptures.* The elders put out a warning concerning his high risk "interpretations" and Jesus said of them, *Beware of false prophets.* We are obliged to ask questions and he is obliged to give a better answer than that.

Jesus has a way of alienating people. *But I say unto you, that whosoever looks on a woman to lust after her hath committed adultery with her already in his heart. And if thy right eye offend thee, pluck it out, and cast it from thee: for it is profitable for thee that one of thy members should perish, and not that thy whole body should be cast into hell.* What can you say to that? The prophet has some strange ideas. People started to wonder if he was becoming a little too intrusive and controlling and exactly when did self-mutilation cease being a good thing. There is a dark side emerging. The messiah candidate is beginning to scare people. If elected, he might demand they pluck out eyes and rational thinking.

There is a siege mentality about him. *Therefore, I say unto you, take no thought for your life, what ye shall eat, or what ye shall drink; nor yet for your body, what ye shall put on. Is not the life more than meat, and the body than raiment?* He lacks real policies to address people's day to day concerns. Instead, he paints a dark future. He is morbid. His dark future is based on his corrosive policies. His idea of society is unworkable. There is a lot of noise but little sense. The message was for the crowd to be glad for the opportunity of dying for him. One thing was clear and that was the lack of direction. The public will learn to despise him. Take his miracle power away from him and you soon see that he is provoking a backlash from the good people of Galilee.

*Thomas the Doubter*

There I was living my contented life in routine service of my master. I had the most enviable of lifestyles because I served in an affluent household. I had a safe and secure place to work and sleep. Every day I could look forward to a meal. Then one day tragedy struck the happy household of Jairus when his only daughter, his twelve-year-old darling, fell seriously ill. We all did what we could do to help but it was of no avail. We prayed ceaselessly that she would get better. But every day it was becoming more and more stark that there was nothing we could do to stave off the inevitable. The master exhausted all medical help and desperation took hold as his whole world lay before him threatening to expire.

So desperate was he that he was willing to go to the enemy for help. You see, Jairus the Pharisee is a prominent figure in the local synagogue, but he was willing to break ranks if it was going to save his girl. It was not the time to care about who was right. There were many good reasons for avoiding the Nazarene and not least to show solidarity with his synagogue. He was driven by his desperate plight. "I will beg for her life" he said as he hurriedly prepared to go. "I know he can save her" he said of his last hope. Suddenly, the man our household had been warned about and told to avoid was the next guest of honour we had to prepare for. Some said his power came from the devil but whatever people said we needed for him to come quickly.

It was never a case of doubting the healing attributed to him. There were too many eyewitnesses and too many testimonies and too many too long in remission for it to be trickery. The problem was his politics. He accused everyone of failure. He was making it difficult for people to like him and many stopped trying. He was left to his own devices but now we had need of one of those devices. An emotional father defied his own suspicions as he threw himself at the feet of the miracle worker and begged for the life of his sweet girl. He said with hope in his heart, *My little daughter lies at the point of death. I pray thee come and lay thy hands on her, that she may be healed; and she shall live.* The controversy and offences ceased to matter.

But as circumstances transpired it also ceased to matter that he could heal. Before my master could get back with the healer it befell me to send out a messenger with the awful news that we had lost her. The Jesus of Nazareth show was no longer required. It was too late to be thinking about mending her. I could not begin to imagine the devastation on receiving news of her death. I could only shed tears for the heartbreak that was his. I waited for the return of a broken man, but word came back that we should keep busy preparing for the visit of the miracle worker. On hearing the tragic news his response to Jairus was to carry on believing. Unbelievable, I thought. It was too late but let's not stop believing in him. It was true he was carried away with believing in himself.

Sometimes all the believing in the world is not as helpful as walking a little faster. I heard Jesus allowed himself to be distracted by the believing woman who stole a miracle off the hem of his garment. Jesus just had to stop and make a show of his incredible powers. Belief and fame seem to go hand in hand, and he is someone who enjoys both. Finally, they arrived with one clinging desperately to the ghost of a hope and the other believing in himself to the point of distraction. The first words I heard him speak I considered to be cruel. He said *Why make ye this ado and weep? The damsel is not dead but sleeps.* If he was for real, he would have known. At this our most tragic hour he was exposed a fraud. He asked that I accompany them into the room where she lay still.

I saw him take her hand and as he did, he uttered words I will never forget, *Damsel, I say to thee, arise.* Once again there was a little girl in our dreary lives. She was up and about like there were things that needed doing. I rubbed my eyes in disbelief. Restored to full health with a simple command like he was God or something! So, it was true what we were hearing about this amazing man but why then was he so unpopular with the rulers? Why was he dismissed out of hand when his gift was so apparent? Surely, the spirit of the Lord is upon him. I couldn't wait to tell someone. But he told us to keep it a secret. Suddenly he was afraid people would find out and start believing in him. This Jesus did not make any sense

*The Apostle Peter*

Elijah arrived later than he was expected. He arrived but only after he was being missed. I asked Jesus *why say the scribes that Elijah must come first?* I had no idea there was a prophecy which was so precise. It is to be found in the book of Malachi. *Behold, I will send you Elijah the prophet before the coming of the great and dreadful day of the LORD.* Finally, they understood him but used what he was saying for another disqualification. They said the Baptist was not Elijah and therefore he was not that LORD. As far as they were concerned, they had found another reason why he was not Yahweh. A few days later Jesus asked me a question. *But whom say ye that I am?* From what I knew I was able to ascertain he was the LORD Christ.

Elijah finally arrived and brought Moses with him. The prophecy was squeezed down to three witnesses. Coming down the mountain James, John and I asked the same question we asked when he stilled the storm. What manner of man is this? We couldn't wait to tell the others. We were stopped in our tracks with the command that we tell no one. It would have to be our secret till the Son-of-man be risen from the dead. Death prophecies were stalking him. The first time Jesus ever spoke to us about dying I promised him I would not let it happen and I felt like reminding him of the marvellous thing just happened. He has God's favour. He is the LORD yet he needs reassuring!

The meeting with Moses and Elijah certainly confirmed his importance in divine plans. I remembered the Baptist saying, *There cometh one mightier than I after me, the latchet of whose shoes I am not worthy to stoop down and unloose.* Nothing could dampen my enthusiasm after my mountain top experience. It takes just one miracle to turn things around. We knew Jesus had that in him. We never considered failure because you do not associate failure with God. The Baptist doubted his messiah, but then he didn't see the Elijah-prophecy fulfilled. I was a part of the same plan. I too was being relied on by heaven. It was my responsibility to get him safely to Jerusalem. We had a prophecy to deliver.

The day came when we were stood on the temple mount. Jesus steeled himself for the task ahead. The next day we let Jerusalem know we were there. The LORD announced his arrival with a riot. The incident ended with him saying, *My house shall be called of all nations the house of prayer.* The temple rulers knew of us by reputation. We are the end-times loving Galileans who are anti-divorce, anti-wealth, anti-progress and of course anti-Roman. Some knew us to be anti-Pharisee, anti-Sadducee and anti-establishment. We are a controversial sect and they found out for themselves. *The people were astonished at his doctrine.* Out of the thousands of new faces going in and out of the temple that day we stood out.

Speaking of revolution, he said *it is written... but I say unto you.* As for all the reasons for rejecting him he said, *You err, not knowing the Scriptures, nor the power of God.* It was the turn of Jerusalem to be left dumbstruck by miracles of healing. Passover was international. There was a new crowd every year. Just as it was in Galilee it started off with excitement. The curious cared not for the inner struggle between the priests and prophets of Israel. It was the temple rulers who needed convincing because they were the ones we had to win over. The messiah-elect was in town for an audience with the Chief High Priest. We needed his endorsement. He could make it official. He therefore also had a veto.

In the Courtyard of the Gentiles there was no call for Gentiles to give up wealth and riches. There was no call for them to take up the cross. There was no attempt to control their lives. Miracles did not depend on them first having faith. The end times was left out of it. It was not for them. He was drawing attention to himself and making himself the subject of his preaching. This, I guess, was always the case but before in such an obvious way. But when he turned to the temple rulers it was to denounce them as hypocrites and blind guides. It was Galilee all over again. By the end of Monday of that Passover we disciples were emotionally drained. We retired to Bethany where we discussed the need for a change of tactics. We were obliged to ask for an audience with the Chief High Priest. We were alienating the ones meant to inaugurate the king.

Tuesday, we disciples were busy making sure his security was not compromised. The wider Galilean community asked why he was not being more conciliatory? We all knew those in charge had to be won over. The alternative was usurpation and a coup. We would be no better than the villains in the parables. It was more of the same tension and animosity. They were not talking to each other. Then finally came a lull in the storm. A delegation of priests came over to investigate the prophetic tradition he represented. It was the nearest thing to an official interview. It probably had something to do with the surge in interest in the Gentile community. They made the first move. They asked, *is it lawful to give tribute to Caesar?*

It was an old question. The rulers came in for unfair criticism and were sometimes accused of collusion when in fact their Pax Romana was a diplomatic success. It was relief for us and a pleasant surprise for them to learn they had his sympathy. He said, *render to Caesar the things that are Caesar's, and to God the things that are God's.* At last, it was friendly, and he seemed to be winning them over. The second question was about the resurrection and again it concluded amicably. Jesus did best on the third. *The first of all the commandments is, Hear, O Israel; The Lord our God is one Lord: And thou shalt love the Lord thy God with all thy heart, and with all thy soul, and with all thy mind, and with all thy strength: this is the first commandment.* They took comfort in the fact that God and Israel are in an exclusive relationship.

The questions were dealt with respectfully. Once more I could hope Jesus of Nazareth could be taken seriously. He could after all impress our leaders. The calm did not last for long. Jesus wanted for the leaders of Israel to know he was God Almighty and not a prophet or the Messiah or the Son of God. He asked a question of his own. *What think ye of messiah? whose son is he? They say unto him, The Son of David. He saith unto them, How then does David in spirit call him Lord, saying, The Lord said unto my Lord, Sit thou on my right hand, till I make thine enemies thy footstool? If David then call him Lord, how is he his son?* They knew what he was about. The one great constancy in his life was his belief in the God of War.

From the age of thirteen Jesus decided to go his own way and favour the God of War over Hellenistic Judaism. He crossed a line when belief in the existence of the God of War turned into him being that self-same God of War. According to prophecy, David's future son of one thousand years is also his contemporary which made his son and their messiah at least one thousand years old. The silence was deafening. You could hear a pin drop. It was broken by the steps of Galileans walking away. It was followed by the push of the Temple Guard. We were escorted to the nearest exit. But it was too late because minds were polluted with blasphemy. Somebody finally said Yahweh and spoke of their transcendent God as if familiar with him. Their cultural defences breached. Their intellectual discipline punctured. Their spiritual purity corrupted. Their philosophy perverted.

The Temple was made to think the unthinkable. The spiritualisation of God was a mistake.

*Claudia the Wife of Pilate*

Every year the Courtyard of the Gentiles attracts visitors from further afield. The grand temple complex and elevated walkways is fast becoming one of the wonders of the modern world. Here you see Rome at its shiny best. Long-term projects like the building of this monument depends on a safe and secure environment. We Romans for decades have provided the international stability needed for prosperity. Passover is the busiest week in the temple calendar and this one-week Romans are invited to share in Jewish celebrations. Passover is the time when Jerusalem blossoms. It is the most colourful of spectacles. For those who want to know more about the Jewish religion there is a synagogue for Gentiles.

We Romans worship our own gods. We have Jupiter, Mars, Vesta, and such like. The Jews do not have traditional gods. They are more like philosophers. Their idea of God is something they call the Supreme Being who is the creator of something they call the universe. We Romans only have our empire because our gods prevail over gods of other nations. As it is in heaven, so it is reflected here on earth. If we start worshipping gods of other nations, we risk our empire. Without Jupiter we would be lost. I so hate it when conceited philosophers say that believing in traditional gods is childish. They think they are so clever. Their creator-God makes them feel more important about themselves. We don't just feel it, we are important.

I am glad to accept their open invitation, but this is the furthest I wish to be from Jupiter. The Temple Guard is out in force today. There is talk of this holy place being violated yesterday and Passover marred by the prophet known as Jesus of Nazareth. Security is increased just to keep him out. He was banned from returning when he was unceremoniously thrown out but that was yesterday. I did manage to see him. There are so many reports of people being healed. I was thinking on this when suddenly he was here in the Courtyard of the Gentiles. I saw him striding up and down the Courtyard as he accused the Jews of betrayal.

It was fortunate for him he had his Greek bodyguard in attendance. To my great surprise, I knew some of them providing him with his human shield. When did Apollos and Stephen become militant? How this unlikely alliance came about I do not know but it was a sight to behold. This Jesus went about provoking the Jews and every time they came to the boil the Greek wall moved forward to form an impenetrable barrier. They were not afraid of the Temple Guard. They were not afraid of a few bruises and a little bleeding. We stayed back and we could see neither side was backing down. Before my very eyes Passover deteriorated into a slanging match between the Jews and their prophet.

His first thunderbolt was *ye both know me and ye know whence I am: and I am not come of myself, but he that sent me is true, whom ye know not. But I know him: for I am from him and he has sent me.* I had no idea what he was talking about, but the Jews had the attention of us Gentiles. *Ye are from beneath; I am from above: ye are of this world; I am not of this world.* I had never heard anything like it. He carried on with his complaint, *if you were Abraham's children, you would do the works of Abraham. But now you seek to kill me, a man that has told you the truth which I have heard of God: this did not Abraham.* Bringing this Abraham into the argument further enraged the Jews. It was a good thing they were not being allowed near him.

This Jesus was having his moment, *Did not Moses give you the law and yet none of you keeps the law? Why go ye about to kill me?* He kept moving about fast. He was like Mars and Mercury rolled into one. He kept disappearing into the Greek wave. The man from above had plenty to get off his chest. *Search the scriptures; for in them ye think ye have eternal life:*

*and they are they which testify of me.* The man from above was also from the past and just in case they had missed it. *Your father Abraham rejoiced to see my day: and he saw it and was glad. Verily, verily, I say unto you, Before Abraham was, I Am.* And the Greeks surged forward to rescue him from a thousand missiles. He was the ruin of Passover.

Thousands of pilgrims have come from faraway places but not to witness this chaos, not to see him. They did not like what he had to say or the way he went about saying it. It is no wonder the temple rulers put a ban on him yesterday, but he managed to slip through to wreak havoc. He diverted attention away from the Jews and the Jewish way of life. Passover became all about him. It made me wonder what Jupiter would do and how he would deal with a similar situation. What if Jupiter was forced to address institutionalised unbelief and sedition? I think I would make the right choice. But what if all Romans were to become philosophers? I would make the same choice.

The man who reminded me of Mars and Mercury was a reminder of what could happen in Rome. His thunderbolts only made Gentiles more curious about Jewish history. Was there a conspiracy against the God or War and did he deserve our sympathy, or should we Romans too be outraged? A time of reckoning had unexpectedly come upon the Jews. He accused them of dirty tricks, and he insisted on washing his dirty laundry in public. The Jews refused to die of embarrassment. Yahweh-Christ refuses to die of embarrassment. They seek to kill him and why wouldn't they. He has a Greek shield protecting him, but they have a Roman one.

*Caiaphas the Chief High Priest*

I do not believe in angels and demons. I do not believe in sky-gods and supernatural creatures. I do not believe in an eternal and external soul. I do not believe in a messiah who is divine or divinely appointed. I believe in the Torah, the five books of Moses. I believe in divine laws that are eternal. I believe in theocracy and Intelligent Design which incorporates divine laws. I believe in free will because I choose what to believe and what not to believe. I believe in enlightenment. I believe that to know of God's existence is the highest virtue attainable. I do not understand the emotional mess which is Jesus of Nazareth. A man with so many random beliefs is hard to keep up with. If I believed half the things this man does my head would be in a spin.

The fact is I am forced to deal with him. He has stolen our proselytes with his idea of a spaceman divinity. He has managed to convince them he is the spaceman. They think they recognise him. They swear allegiance and he is emboldened by them. Moses was clear, *God is not a man* and *neither the son of man*. Jesus of Nazareth is in breach of the Torah. The man is desperate for attention. Sensationalism gives it to him. There is no reason to interview him because we already know what he is about. He can break all the commandments he wants but it will not be granted. By his own admission he is a man and with that we agree. He said, *...me, a man that has told you the truth*. Each time we expose a flaw he moves on to the next blasphemy.

It is Scripture which condemns him. Everyone in Nazareth knows when, where and to whom he was born. The townsfolk could put up with him having his personal interpretation, but they really did not like being implicated in his attempt to swindle the prophets. They knew him well enough to want to kill him. I said to Nicodemus, *are you also of Galilee? Search and look: for out of Galilee arises no prophet.* One moment he is a man and the son of David. The next time you see him he is the spaceman and the Lord of David. He wants our help to steal the throne of David. We want it on record that he failed to deceive us. He accused us of error and that charge is escalated to usurping the prophets.

We frustrated his campaign by informing the people of the contents of scripture. Instead of accepting our decision he further twists prophecy to make it all about something else. He is one of us and we have done our best to accommodate him. We left him to his own devices waiting for him to turn the corner. Now he brings our proselytes into it. The time is come to tackle the problem head on. Meanwhile, we reminded the Greek synagogue that God is not a man and a year ago that was what they believed. The true idea of God is of pure God, and he is not that. God is not a part of creation or part creation. Moses resorted to anthropomorphism to convey a complex idea to those who did not have his education.

I guess there will always be someone out there with a literal interpretation of the Old Testament. Either that person foolishly believes a creature of the universe created the universe or he does not care because what matters to him is that he is a prophet. Sometimes faith compensates for lack of knowledge and sometimes for common sense. But we go by scripture which clearly states Bethlehem so why must I waste my time on this fraud from Nazareth? We know the law and we know he has a grasp of basic concepts. We know where each other stands. It would not surprise me that when he said *you seek to kill me,* he was inviting arrest. One way or another he intends to get an audience. He is that desperate for attention. He is another Socrates.

Exactly what is it that he expects from me? No one is disputing the fact that blind eyes are opened but what can I do for him? We will not be bringing upon ourselves the end-times, as he calls it. We don't want Armageddon. We want our civilisation to continue to flourish. If the Sanhedrin did comply with the wishes of the hermit, the random messiah will have to administer justice and keep the peace. He will have to make difficult decisions. Is he ready for that kind of responsibility? Is he really suggesting we give up other options and simply take up the cross? We who do not want war and have done everything to avoid it. I will not be sending my children to war just for the pleasure of calling him Majesty. His truth is selfish, violent, and ugly.

Truly, it is his disciples who have had the most absurd experience of it. They have been called upon to reject the very faith by which they concluded he was messiah. Meanwhile, we have had the pleasure of teaching him a very important lesson. You cannot always have what you want. Now we need to teach him that you can end up losing what you have already got. He is excluded from Passover. We will catch up with him and give him his much sought-after audience before the judges of Israel. He will pay for his blasphemies. His is brazen. He used the issue of blasphemy as an excuse for further blasphemy. He said, *say you of him whom the Father has sent into the world, Thou blaspheme; because I said I am the Son of God?*

He is wanted but not in the way he hoped for. He does not know how lucky he is. We gave him a home and family. We gave him Jewish citizenship and shielded from the attention and interference of Rome. We gave him an identity, a life. It shows because we are a major part of his consciousness which is why he is not able to leave us alone. He needs to prove himself to us. He keeps coming back demanding to be taken seriously. He has set off a chain reaction and we watch the momentum build. He has set off a loathing. Groups of young men are prowling the streets looking to vindicate Moses in their own way. I look forward to meeting the man who threatened to give our vineyard away to others.

He will find out just how much we mean for his wellbeing when he loses his Jewish citizenship.

*Cornelius the Centurion*

I am telling my story because the truth matters. I don't know why things happened as they did, but I know they happened for a fact. I want to give an honest account, one without fear or favor. My commission took me to Galilee and while there my loyal servant fell ill. His condition worsened and I feared for his life. Doctors attended to him but there was no improvement. His situation became desperate. The growing reputation of a miracle worker reached my door. But every time I thought about going to him, common sense talked me out of it. Miracles just don't happen. Stories of remarkable healing increased and naturally I had to know as much as possible about him.

My source, a Jewish official, told me about prophets. A selection of notable individuals who believed they were called into service by God. They acted as intermediary between the God of Israel and Israel. One thing which set them apart was miracles. I was told the story of Naaman the Syrian General. He was a Gentile like me, and he suffered from leprosy. He too heard of a prophet with healing powers. But there was a likelihood the prophet of Israel might not want to be associated with a worshipper of Rimmon. The prophet of Israel was under no obligation to help him. Naaman the leper went to him out of desperation. At first, the prophet Elisha refused to see him. As if to add insult to injury he sent an underling to tell the mighty General to go and wash in a river.

The General struggled with the random and improbable nature of the treatment but eventually decided the chance of being cured was worth the risk of making a fool of himself. Naaman overcame his pride and did as he was told, he humbled himself. The first Gentile to get anywhere near the God of Israel had his health restored. The First Gentile even managed to impress the God of Israel. Naaman said about his experience, *Behold, now I know that there is no God in all the earth, but in Israel.* He was a monotheist before there was monotheism. The Syrian was remarkable for his faith in the God of Israel who said to him, *Go in peace.* I was ready to go and find my prophet.

I caught up with him in Capernaum and it was just after he healed a man with leprosy. I felt like I was taking a turn after Naaman. The prophet seemed to know I had no interest in the nationalism he was preaching. That was a concern for King Herod. Neither of us had any ideas of supremacy. He was working for Israel, and I could respect that. My concern was to be ready to be tested. I was ready to show the same humility as Naaman. I explained the situation which brought me to him. My servant was on his deathbed. It took me by surprise that the prophet would offer to come to my house. He was no Elisha. This was not a person who was standoffish. He was willing to set his reputation aside. But time was against us. I said to him, *Lord, I am not worthy that you should come under my roof: but speak the word only and my servant shall be healed.*

I went home and found my beloved servant restored to good health. I did not see the prophet again till the Jewish Passover. He was under arrest and all Jerusalem was talking about him. Members of the Greek synagogue were on hand to explain everything to Gentiles. One told me the story of Yahweh the God of War. To my surprise it was a familiar one because the same was happening to Mars the Avenger and Roman God of War. Sky-gods were under threat of the Supreme Being of Philosophy and the natural sciences. Spirituality was a cultural change we were all aware of. The situation was a little different for Yahweh, he was already rubbed out of history. The story was that he had emerged out of the Word of God to fight his corner. The story was that of the Greek Christians.

He was charged with sedition, given a fair trial, found guilty and the conviction was sealed with crucifixion. Israel rejected him but it was also Rome's prerogative to make kings. Pilate did not crucify him because his name was Yahweh. Others called him Yahweh but that was the crime of others. It was only a crime against Rome if he was Yahweh. The Roman Emperor is God manifest and the empire has only one. A second would be a challenge to Rome's supremacy. Rome's supremacy was maintained. Meanwhile, the Greek contingent got what they pressured for. It was a fair trial and Pontius Pilate had followed due process. The judgement of the court meant the offender had the recognition he craved, and it was official and legal. For the remaining hours of his life, it was lawful to call him Yahweh.

I am the one who after watching him die proclaimed, *Surely, this man was the Son of God.* He did not wait to die. He set himself free with a defiant scream. The fact that he cheated the cross is a story I will be telling for years to come. His last miracle robbed the cross of its victim. When I told the Governor he was gone Pilate could not believe a death so premature. We did not kill him. We only dictated where he should die. By taking back control he managed to have the last word on the matter. He turned his death into ceremony. He used Rome to get his name back. Even from beyond the grave he was able to manipulate the philosophers. At the behest of the priests, we Romans ended up protecting his body.

# Chapter 7

# The God of War

I watched avidly as the Oxford University don presented his documentary "Root of All Evil?" It was sensational stuff because it was the first time on British television religion was attacked the way it was. I was expecting your usual cool headed intellectual discourse but to my surprise the programme was not fettered by the habitual and obligatory respect for religion. This was something new. It was full of passion and anger. There was confrontation and courage as the brave academic showed belligerence where I expected to see the usual deference to religious leaders. The argumentative professor was providing opposition and not scepticism. In 2006 militant atheism made its way into my life in the form of an evolutionary biologist called Richard Dawkins.

I will never forget the moment he introduced Yahweh. With glee he listed the thuggish characteristics of the personality god. His description of the violent God of War was reason enough for rejecting the deity. He was describing the God of Ancient Israel. The Bible as appreciated by Dawkins was far too literal and selective for my liking. But where was he going with it? This great and self-absorbed mind of our times was condescending whenever talking about Yahweh. But then any teacher type is going to feel superior when confronted with such a childish concept of divinity. I thought everyone knew the spaceman was invented to express the difficult idea of God to little children. God is to be feared and Yahweh achieves that. The intellectual seemed not to recognise a literary device. The skygod is a metaphor. God proper is a complex concept.

I went along with it. After the character assassination I found myself still in the zone. My first response was one of pity for the skygod. It is no wonder Yahweh is always angry. It is because absolutely no one is taking him seriously. I wondered if I could change that. I identified the need for someone to take the monster seriously. I felt my Christian past qualified me. The groundwork undertaken by Dawkins meant the hardest part, separating the God of the Jews from the Jews was already done for me. I had Yahweh handed to me on a plate. Dawkins introduced an unpleasant character which meant I had to represent him as someone misunderstood. My curiosity turned into empathy. It said something about me. I too was someone who wanted to be taken seriously.

I decided to ignore all the "skygod" objections and like Dawkins engage the Bible afresh. His intellectual arrogance had rubbed off onto me. Yahweh the comic strip character certainly helped gain traction into a new world. I had no problem assimilating the cult of Yahweh. I caught up with Dawkins. The Word of God is otherwise a sophisticated mix of theism, mythology, fable, and history. Religious principles and concepts are dramatized with the use of different literary styles. The story of Adam and Eve is an origins story. The creation of man asserts a moral being. The story introduces sin and with it the Bible acquires an overarching theme. It is not a bad effort by the ancients. Everyone from every culture has a right to some sort of understanding of their world. It is standout literature but that does not make it the Word of God.

The compilation of books known as the Old Testament came by way of editorial process. Faith is an editorial tool. It takes faith to make sense of the styles and content. For most of us this means accepting the terms and conditions of the Christian church. Like Dawkins and thousands of others I once read the Bible using their faith. But as a lapsed Christian I could easily suspend faith and appreciate a faithless rendition of the Bible. Christianity without theology is the personality cult of Jesus Christ. The Old Testament without theology is the cult of Yahweh. I agreed with the assessment of Yahweh's immature character. He is reactive. He is emotionally raw. He gets angry and ends up killing people. He is a moral contaminant and not the sort of company a reasonable person would want to keep. He is a control freak. It hides his inadequacies.

Although Dawkins read the Bible with an open mind the militant atheist is opinionated. Still, it was his scientific discipline which had me listening. He was trying to read the Bible without Christian prejudice and it was showing. A god in the image of man is usually made by man but he persevered. He came up with a personality profile. His discipline yielded an all-important personality profile. I was prompted to ask, what is the story with Yahweh? Every reader must decide if to go with Yahweh the Creator or Yahweh the spaceman. With so many ideas in and outside the Bible it is left for church leaders to authorise beliefs. Orthodoxy is the belief of church leaders. They favour the primary substance. Dawkins was trying to embarrass them with the spaceman.

A second tier of beliefs to that of orthodoxy is needed by those exposed to evolution and Big Bang. Personal faith is different for everyone because it is personal and the personal can mean freakishly different because some people are. Orthodoxy, for me, was the belief system held by people I wanted to socialise with. I adopted orthodoxy to get in with the in-crowd. Dawkins was about something else. He wanted to prove that all spiritual experiences are a matter of delusion. He kindly provided definitions and explanations. He said about the primary substance, "The reality we inhabit also contains a supernatural agent who designed the universe and maintains it". This is theism, creationism, and spirituality and it all came out of the primary substance.

Russian Christians, my favourite mortal enemy, think the universe is a spiritual phenomenon and faith is the operating system. Dawkins thinks science is the operating system. Meanwhile, I think the universe is an existential phenomenon and existentialism is the operating system. Existentialism goes best with Big Bang cosmology if it is consistency that you require. Russian Christians believe in a readymade universe. Dawkins was using the spaceman to puncture a hole in that universe. To the annoyance of primary substance practitioners, he refused to take the leap of faith leading to the supernatural agent of creation. Dawkins stopped at Yahweh the spaceman. This was seismic. Dawkins was not interested in what Christianity made of the Bible. He wanted to know what he himself made of it.

The idea of being circumvented came as a great shock to Christians. The paranoid leaders of the Russian Orthodox Church felt more threatened than usual. Dawkins showed a certain lack of courtesy by refusing their gift of faith. He was going it alone. The militant atheist refused to concede to their demand that he be sensible and adopt faith. Thousands of years of theology was going to waste. There was a time he would have been hauled into a religious court and made to pay for his pride. But it was too late for the faithful. It was out of their control and there was nothing they could do about it. In 2006 the world was introduced to Yahweh the spaceman and God of War. Dawkins exposed the Bible's "inner contradiction". Except it was a faith conspiracy.

Dawkins the inquisitor was tearing into creationists because they banned the teaching of evolution in schools. It was his subject and that made it personal. But the real springboard which explained his animosity toward religious leaders was the terrorist attack of 9/11. He was outraged. He experienced the same mix of feelings Jean-Paul Sartre felt after World War 2. Something had to be done about it. Faith fascism like Nazi fascism declared outright war on secular society. When division is of such extreme nature it is time to say we owe them nothing. There must be no goodwill toward fascists. Christians are not nice fascists because there is no such thing as a nice fascist. Theocracy is totalitarianism and Christians are to one degree or another theologian. Once again, the primary substance needed to be made safe. Once again, a biologist stepped forward.

In 2006 the delusion could not be allowed to go on because it was leading to World War 3. History was repeating itself. The authority figure was angry and every moment on screen he was becoming more like Yahweh the God of War. Wronged and vindictive and vengeful was there for all to see. The Big Bang theory and the theory of Evolution cannot be undone. Progress cannot be reversed for the sake of the deluded. He embodied science and science was furious. The spiritual conspiracy was against him too. They said he had a devil in him. They did not want to hear what he had to say because everything he said was anti-faith and anti-primary substance. Their heads were stuffed with nonsense, and they did not have ears to hear. It did not take much of Dawkins' thinking capacity to work out the trajectory faith was on.

In 1945 the nations of Europe turned the corner on Year Zero. In Paris a new creed was devised the purpose of which was to erode fascist methodology. Equally anti-faith and anti-church existentialism was unleashed. Faith-heads no longer had it all their own way. In 2006 the creed manifested as militant atheism. The only thing left to do was to create an existentialist religion with an existentialist God. It would truly mean an end to all the suppression, oppression, coercion, and extortion. Church law is a thing of the past and it must remain there. Blasphemy in Russia leads to a criminal record, fines, imprisonment, and ostracism. The future is bleak for anyone with a criminal record. But back to Dawkins the evolutionary biologist. Evolution was first to contradict church teaching and render obsolete all forms of spirituality.

If we are wholly biological, then spirituality has no relevance for us. Teaching evolution is cultural suicide for primary substance practitioners. Evolutionary Biology threatens to change man from a spirit-being to a genetic one. Meanwhile, Bible belt evangelicals insist the Bible is the Word of God. Therefore, man was made readymade. Modern readymade creation comes with the proviso that the illusion of evolution is built into it. Evolution was designed to be a test of faith. Everything which questions the nonsense is a test of faith. As far as faith-heads are concerned scientists are allowing themselves to be deceived. They have failed the test of faith. It is only Yahweh the God of War who can save us from faith and the formidable faith family.

In 1945 fascism was defeated. The new world outside of churches was a changing world. Christianity was left to its own devices. Christians needed to pray about any proposed changes to their divine constitution. In 2006 prayer was interrupted by a militant atheist. He said faith is a system that has been designed to deceive itself. It is a system for sustaining the lie on which it is built and that being the primary substance. Christians said they would pray about that too. They were still praying in 2010 when Stephen Hawking lost patience and declared philosophy dead. In all, it took sixty-five years for existentialism to triumph over faith. There were pockets of resistance. Totalitarian forces in Russia combined to keep fascist methods operational. Kremlin Christianity was born.

Dawkins did his best to pre-empt World War 3. His best weapon against the God of the Christians was the God of War.

But the spaceman idea never got off the ground. No one knew how to use the idea or what to make of the spaceman. Dawkins was at first embarrassed by the spaceman deity. But so also were the Bible thumpers. An opportunity was staring him in the face, and he tried to make the most of it. A skygod is different and there are hundreds of them and each with their own remit. No one looked to Neptune for an idea of how the world had been created. Roman soldiers kept praying to Mars for protection during battles. The primary substance satisfied intellectual curiosity for an emerging intellectual elite. If they prayed for anything it was for answers.

At first there was no competition between the gods. Yahweh remained the Israelite God of War and enemy of Gentiles. No one cringed when reading about God and Jacob wrestling. Competition for students came to temple cults and schools of philosophy. Priests and professors vied for the best students. Skygods were relegated to secondary gods. The primary substance was deemed an imaginary friend. Students and apprentices were denied neither education nor traditional worship. Society progressed. Neither Dawkins nor the patriarch of the Russian Orthodox Church can explain the God of War. Neither are Israelite and so they do not matter, and it does not matter what they think. They don't count. Aristotle, a non-Jew, opened his own school of philosophy where he taught the natural sciences and natural theology. His experiments showed processes at work, all attributed to the primary substance.

Scientists were obsessed with finding out more about the processes. What were previously secret and hidden processes were revealed. They suggested intelligent design. There was an intelligent creative agent at work in nature. It knew what was required and it knew what to do. A creator in principle or God in principle drew Greeks away from the traditional gods. Science was preferred to mythology. Yahweh the God of War is Jewish mythology. Dawkins ended up going to Yahweh's emotional rescue. He made the spaceman world famous in 2006. Evolution is not the only blasphemy. He had millions of people the world over saying Yahweh. I was one of them.

The passion of Richard Dawkins as seen on television was not dissimilar to that of Christ's passion. Both parties were surrounded and roused by theologians. Both said of creationists, they do not have ears to hear. It is no wonder Dawkins was always on edge. But imagine what it was like for the spaceman having to battle the same faith family. Being passionate or emotional is being true to yourself. Emotional responses vary from person to person. Responses vary for the same person at different stages of life and development. The passion of Richard Dawkins like that of Yahweh's was to inform the Jews of the delusion they had fallen into. It was to restore consciousness of the God of War. The Professor for the Public Understanding of Science was pleading for Religion to stop suffocating science. For the God of War, it was a little more personal.

Plato the theist said, "the knowledge at which geometry aims is knowledge of the eternal". Philosophers were in awe of themselves and rightly so. They were crusading academics bringing higher education to the world. They had intellectual discipline and were a new kind of elite. They were about the natural world, and they invented universal truth. Their idea of eternal was the science underlying the natural world. They thought in terms of primary substance. Intimate knowledge of the universe was the then equivalent of a personal relationship with the creator. Intelligent Design meant they had found the god they themselves had invented. In turn, the world as they understood it justified their philosophy. They believed they were looking into the mind of God.

We would fault Plato's school for making some big assumptions. But philosophers followed the science, and the same people would be doing that today. They would change assumptions to get to the truth. Pontius Pilate had to look into the eyes of Jesus Christ to see into the mind of God. Dawkins the avenger read the Bible for insight. He wanted to debate Evolution with Christians and so it was in his interest to have some idea of what they believed and why. To his surprise, there was no basis for scientific debate. All he found were stories about a God of War. According to the Bible, the universe was created by a violent and erratic God of War. The militant atheist had no interest in substituting one idea of God for another. He had no intention of defending one idea of God above another.

Dawkins was hoping to find out about the Christian faith. Faith needed fixing. The scientist discovered faith itself is the error. The Word of God is not the Word of God because that mute all pervasive and multi-galactic god of theirs does not exist. The prolific writer knew all about literary devices. The evolutionist put the 2 gods in order of appearance. His book about the method of creation ended up a book about the god delusion. It was not like he had any real interest in the spaceman. But there it was, an anomaly in the faith universe. Dawkins was not interested in getting to the bottom of it. He did not want to think about the merits of a God of War. He was in an anti-faith frame of mind because faith is the delusion. The spaceman is the elephant in the room.

You don't have to be as clever as Dawkins to find the spaceman tucked away in all the lies. But you do have to be as courageous as him if you wish to tackle the creationists over their poor comprehension skills. They institutionalised faith and that has not done anybody any favours. They made a liar of the young Dawkins, but that endeavour was doomed to failure. It turns out that for a lot of Christians faith might as well be called doubt. Seriously, how can anyone believe in a Jesus bigger than a London double-decker bus. It sort of catches up with you. Their crime is against God, man, and universe. Without spirituality, without any of the symptoms Dawkins could only read about God in real-time. He explained real-time.

God told Abraham to kill his son and at that precise moment in time it was exactly what God wanted. We can only imagine the jealous God was jealous of all attention and affection Isaac was receiving. The instruction to murder was not a test. Isaac was in real jeopardy. There was no prophetic relevance to the situation. It was not being acted out for future generations as a lesson in protocol. Abraham knew he had no choice. God wanted the sacrifice. He was being bullied. Real-time meant God really wanted Isaac dead. When he commanded it, he had no intention of changing his mind. Our impression of God changes when the story is put into real-time. This God of the Bible is to be feared because he is without conscience. The Ancient Israelites were warned to fear the Lord and now we know why.

Yahweh has no conscience, but his conscience found a way around that little problem. It was his conscience which went beyond the original impulse and whim. Conscience was able to correct self. He too is a double entity, but his conscience is not like ours. Our conscience is in the same body and is the same body. Conscience acts and intervenes when it suits conscience. God had to learn to be jealous. God is not an infinite continuum and there is no such thing as continuum jumping or prophecy. With the God of War what you see is what you get and what you get is evolving. The universe has a beginning. It is finite and it is expanding. It is sustained or kept from collapsing by the force of expansion. It is not being kept in place by the primary substance or any other philosophical entity like the Christian logos.

The universe is a physical phenomenon. It is the physics of matter which allows for it to be traced back 13.7 billion years to the hot dense primordial soup of matter. The evolutionary biologist was joining up the dots to the beginning of life on Earth. His personality profile of Yahweh revealed an existential person. It seems to be that *The Selfish Gene* was the first dot in his journey in more ways than one. The geopolitical God of Israel is jealous and selfish. The scientist has taken existentialism from philosophical to scientific. He has corrected existentialism. It is not the case that existence precedes essence. Existence and essence come together as one in the selfish gene.

The biologist read the Bible and was surprised by the biological concept of God that it had. He tried to ignore it. He said, "*I know you don't believe in an old bearded man sitting on a cloud so let's not waste any more time on that.*" After exposure to the faith family, he entered a marriage of convenience with the God of War. The family believed in the strangest of things and it was because of faith. They would not listen to reason, and it was because of faith. Dawkins weaponised his find and ended up becoming the first person in history to take Yahweh seriously. He was angry and not in any mood to grant concessions to theologians. The fact that he was free of theology meant he could take the God of War seriously. The fact that he was at war with faith meant he had to.

The faith family is a wicked conspiracy. When over one billion people agree on the existence of demons there is not a lot any sane person can do about it. My favourite gripe is them relabelling spirituality as morality. Kremlin Christianity is engaged in de-Satanizing Ukraine. Let us never forget that nuclear Christians worship a Jewish man and not the God of War. Faith like propaganda cannot correct itself. Every new diktat brings confusion to the last one. Any correction must come from outside the system because the system is built to deceive itself. Falsehoods come as second nature. Navalny is outside the system. He knows the difference between a real war and a patriotic war. Kremlin thugs invented the patriotic war to stay in power. Navalny experienced hunger many times over during the cold war. Anaximander made a mistake and Russia is without freedom of speech.

By taking the God of the Old Testament seriously Dawkins was able to attribute meaninglessness to God and the Old Testament. The God of Israel is for Israel, lately the Jews. The relationship is exclusive. He is nothing to do with Gentiles. I am a Gentile, and he would prefer I stay out of his business. There is no urgency on the part of the God of the Old Testament to save me from sin. Anyway, he gave his commandments to Israel and only they can be in breach of his commandments. The God of Israel and the Old Testament should inspire meaninglessness in Gentiles. Dawkins created his very own God of Meaninglessness in the form of the Flying Spaghetti Monster. Theology raised expectations. Servile temple cults needed modernising or risked becoming irrelevant.

The scribes could not hang around all day waiting for the God of War to turn up and issue instructions as to what next. Ancient and modern is captured in the Old Testament. Those who wanted tradition could read Yahweh the God of War into the book. Those who wanted to be modern could read Yahweh the creator into the same verses. Readers decided interpretation for themselves. Jesus of Nazareth claimed the spaceman was the correct concept of God. Dawkins followed suit and described the spaceman as the God of the Old Testament. But Jesus went further when he admitted to being the spaceman. Again, there was laughter at the idea that God is a man.

# Chapter 8

# Philosophy is Dead

This is an appeal for help. God needs your help. Someone stole his identity. He wants his name back. Will you help? He reported the crime but making a record of the felony was never going to be enough. The robbers are holding onto his name as tightly as ever. God needs you because only you can get his name back for him. He is desperate and will reward those who rally. First the Jewish and then the Christian theocracy robbed him of his identity. Theocracy is government by priests of one kind or another. The crime was brilliantly executed and successfully covered up and there was no one left to suspect the world's most audacious theft. God himself had to report it. Developments in forensics led to a reopening of the case. In 2016 detective Dan Barker published his case notes in *God: The Most Unpleasant Character in All Fiction.*

Barker stumbled onto something hugely significant. He read the gospels for the hundredth time and concluded Yahweh-Christ. Yahweh and Christ are the same person. Barker at the behest of his mentor Richard Dawkins was writing the follow-up to *The God Delusion.* Dawkins' bestseller revealed an extremely unpleasant side to God. But that was not a reason for denying him justice. Barker used the same methods as his mentor. The ex-evangelist applied real-time to the gospels. The ex-preacher discovered Yahweh hidden in plain sight and the main thrust of his argument was that Christ said he was Yahweh. It was blasphemy for anyone to speak that name and that helped hide the theft of his name. Blasphemy laws made God a criminal. Those laws were repealed in 2006.

For so long the case could not be discussed. It was also a crime of blasphemy to elicit a "Yahweh" from a third party. This explained why the original notes were written up in the way they were. But it all changed in 2006. Meanwhile, there were many advances made in the sciences. By 2016 the changes in law and forensics caught up with the ex-missionary turned detective. Barker realised Christ was blurred by prophecy and legends from the times were included for the readers of those times. Barker matched the intellectual arrogance of his mentor. He decided to put Yahweh-Christ into real-time. A different character emerged out of the same pages. He found the real-time Christ who reprimanded theologians demanding they restore consciousness of Yahweh the God of War.

Barker found Yahweh-Christ in the throes of an existential crisis. What was in heaven real-time became on earth critical time for Yahweh-Christ. Critical time as opposed to prophecy focuses a person's mind on his or her immediate situation. Christ got into many unsafe situations. There was more than one attempt to stone him and sometimes all twelve bodyguards were needed for his protection. They enabled him. After a lifetime of Christianity Barker finally happened upon a reliable fact about Jesus of Nazareth. Christ was suffering from a Yahweh complex. Atheists believe Yahweh is myth but that is not the point. If Richard Dawkins got it right about the spaceman, then it is not so farfetched that Christ thought he was the spaceman.

The gospels after being subjected to unrelenting criticism yielded up Yahweh-Christ. Remarkably, this was not possible before 2006. Of late, there is a new way of reading. Not reading the Bible but reading the sensational Yahweh. Unbeknown to Dawkins he started something. His famous list of unpleasant characteristics brought Yahweh to life. He is someone we can all loathe. The concept of Yahweh-Christ is even newer and just as sensational. The nondescript Christ was the nondescript Yahweh but by 2016 there is fire in his belly. Imagine Mr Nasty as Mr Nice. Imagine a Yahweh approachable and amenable. Imagine Jesus Christ as vengeful and vindictive. Imagine him a God of War. The stunning Yahweh in the gospels changes the gospel story drastically.

Imagine God on a level playing field. Imagine underneath the unpleasant exterior someone trying to right a wrong and desperate to be listened to. The story of God gets a lot more interesting. Dawkins identified the God of War as the God of the Old Testament. The thing being said is the God of War is God. Barker identified the God of War in the gospels. The thing being said is that Christ is God. With Yahweh-Christ we double our information on God. We have the background story the gospels are lacking. We can add to Dawkins' personality profile. We have the story of Yahweh's coming out. The God of War did not become a man. It was not incarnation of the spaceman. What happened was Yahweh became Yahweh 2.0.

There is a familiar problem with the new concept of God. Jesus of Nazareth could not have created the universe. Man arrives too late in the order of the cosmos to even be considered. Jesus of Nazareth cannot be a creature of the universe and its creator at the same time. Never mind blurred, it is absurd. But what then is the point of God if not creator? Anyway, the biological Yahweh was eventually surrendered to divine philosophy which was the cosmology of the day. If we are to keep Yahweh-Christ, then we need to get rid of the silly idea that he created the universe. We need the correct intellectual tools which will allow us to consider the fact that God is a man.

Yahweh and Christ brought together by Barker results in a fuller picture of the entity. But we risk upsetting theologians. Nobody wants the Russian Orthodox Church on their tail. Especially now they have incorporated the Novichok Gang. Dawkins dared to come between Israel and the God of Israel. Barker dared to come between Christ and the Christians. When we make it about the spaceman, we dispose of theology. We lose excess baggage. The New Testament thins out. The gospels thin out. Jesus of Nazareth dared to come between the primary substance and primary substance practitioners. Theologians tried to put a lid on him, but he managed to get past them to attend International Week in Jerusalem.

Every Jew except one attending that Passover was proud to be Jewish. The world's first socio-spiritual centre was enjoying a golden age and there was a lot of wealth passing through the temple coffers. They were forced to pay tribute to Rome but there was a huge dividend. They had brilliant negotiators because they ended up with an empire of their own, a spiritual empire. There were synagogues in every major city. Rich Jews contributed to the temple project bringing it closer to completion. The temple was a source of national pride. Costs had to be absorbed. There was an army of priests and a professional army in the Temple Guard. Welfare projects across Judea and Galilee resulted in social cohesion.

Rendering unto Caesar the things which are Caesars was not without benefit. It was because the temple rulers never lost sight of nationhood. It was what they were about. The temple provided national focus and a yearly measure of progress. Passover was a time of self-congratulating. The temple rulers must have been doing something right. There was even a Greek synagogue in Jerusalem. At Passover the temple's Courtyard of the Gentiles was always packed. Unfinished or the always new House of God was an architectural wonder of international fame. Every year tourist numbers went up. The temple was a symbol of their success.

Then came the year of Yahweh the intruder. He did his best to come between the people and their spiritual leaders. He tested the loyalty the rulers could command. The absent landlord had been absent a long time and that did not work in his favour. He was forgotten. Nothing is a better measure of the Temple's success than the fact that the God of War himself was not able to fracture the nation. Dan Barker found Yahweh-Christ in the Jerusalem temple and that was just about all that could be said about the great and terrible day of the Lord. He was evicted from the temple. They could not tell if it was a threat or a promise when he said *Behold, your house is left unto you desolate.*

Yahweh-Christ stood as much chance against faith as Navalny does against propaganda. He chose to fight a losing battle. Navalny chose to stay when he could have left Russia. Today he languishes in prison. Faith and propaganda are two weapons existentialists fear the most. Each generates a family of fools who will believe anything if it is told repeatedly. Christ was fighting the ghost of Anaximander and Navalny the ghost of Stalin. For a patriotic Russian it is not enough to be a Russian. A patriotic Russian is a supremacist, a crook, a thief, and a liar. I'm just repeating what I have heard. The Soviet Union failed because it was being run by patriotic Russians and the Iron Curtain put an end to their invasions. World War Three might yet be their parting gift.

Back to faith, militant atheists won their war against the theologians. Science was on their side but instead of sweeping away faith it led to polarisation. Primary substance practitioners were satisfied that spirituality and faith had withstood the onslaught of science and scepticism. They celebrated delusion. The atheist army turned its attention to Jesus Christ. They plotted to reduce the probability of his existence. Thank God they did. The gospels have a faith agenda and are about as useful as the Old Testament. Faith was undone and removed from the gospels. This task was undertaken by ex-evangelist Dan Barker. With faith and prophecy removed and the story put into real-time the new conclusion was that Christ said he was Yahweh.

Only a literary invention would admit to being a mythical being. Messiah is an invention of prophecy. Christ is a spiritual and philosophical invention required to unmute the primary substance. The God of War is the anthropomorphic version of Yahweh the creator God. There does not have to be a real person. If Jesus Christ said he was the God of War, then the suggestion is that Christ is another literary invention. He is likely to be another philosophical entity. Artificial and contrived to keep theologians in power and in the money. Strictly speaking, Christ of the gospels is a fictional character translating and casting the theology of the Paul into a true to life setting. But is Paul's Christ based on a real person who lived in the reign of Tiberius.

Paul dismissed Christ after the flesh claiming Christians were better off without God as he really is. He could imagine a better religion and with it a new idea of Christ and God. Hellenistic Judaism had the best idea of God and Paul was a Hellenistic Jew. Social engineering was not beyond him. The God of War had his revenge and served out meaninglessness to the Jews. He said goodbye with, *Behold your house is left to you desolate.* Paul resisted meaninglessness. Dan Barker the ex-preacher wanted to go further with the claim that there wasn't even a *Christ after the flesh.* Whatever his motive he gave us Yahweh-Christ a personality God based on Dawkins' version of Yahweh.

The only reason we can say Yahweh is Richard Dawkins. He fought the God delusion using every means available. He returned God's name to the rightful owner. It took 10 years to get from his Yahweh to Yahweh-Christ. Dan Barker had the intellectual tools and discipline necessary to quash Christianity. He was so angry he was willing to read the revisionist gospels one more time. They made a liar out of him and left him drowning in delusion. They gave him a false world and a false reality. They took away truth and stuffed him with faith. They told him philosophical speculation was truth. They abused his mind. The born-again Christian turned atheist was able to read the gospels and suspend theology at the same time.

Christianity is another religion going all the way back to Anaximander. Christians live in denial because the primary substance has two origins stories. Hundreds of years before Anaximander there was King David (c 1000 BC) who wrote a Psalm which is much loved by theologians. *Where can I go from Your Spirit? Or where can I flee from Your presence? If I ascend into heaven, thou art there: if I make my bed in hell, behold, thou art there. If I take the wings of the morning, and dwell in the uttermost parts of the sea; Even there shall thy hand lead me, and thy right hand shall hold me.* Divine Omnipresence is suggested. The Psalmist was being poetic. We know because Jesus of Nazareth did not possess a detachable hand. The God of War would have said so when he denied being the Messiah. *David therefore himself calleth him Lord; and whence is he then his son?*

By the time of Jesus of Nazareth any suggestion of the spaceman existing was accompanied with howls of laughter. The idea of a spaceman Supreme Being filled creationists with amusement. Not unlike Dawkins reliable intellectuals of the 1st Century also thought it silliness. The world had changed on the God of War. There was a new concept of God which brought with it new "divine" attributes. The God of War could not compete with the Creator God. Barker is an atheist and for him any revival of the cult of Yahweh came from one of many sects. If Yahweh does not exist, then Yahweh-Christ is made up. If he does exist, then he worked out his existential crisis and is once again a public relations disaster.

Dan Barker had to unravel a conspiracy to get to Yahweh-Christ. But is it all conspiracy. The leader of the Christians died on the cross but there was no agreement as to why. Some went for the spiritual theme and that he had died for the sins of the world. Some went for the politico-religious option claiming he died to be recognised the Messiah and King of the Jews. Roman law and Roman punishment said he died for sedition and for challenging the supremacy of Rome. His was a famous death because of who he was. People were soon marvelling over the fact that he had managed to make a case for skygods and in all places it was in Jerusalem, the seat of creationist worship. He won his case against the creationists. Hellenistic Judaism was given a bloody nose. Skygod worshippers acquired a champion.

The Bible's inner contradiction was broken. The God of War broke out of Holy Scripture. The God of the Old Testament walked out of the Old Testament and into Galilee. The House of the Lord was left desolate and so also the Word of God. *Behold, your house and your book are left unto you desolate.* The apostle Paul and other Jews of the Diaspora refused to accept God's resignation. They did everything they could to stop the Greek super apostles acquiring the God of Israel. They did not like desolate and meaninglessness. The Old Testament was of no significance to Gentiles. Any meaningful bits came and went with Yahweh. Christ referred to Moses the person. Gentiles did not receive instruction in Jewish laws and customs.

The God of War was at war with Hellenistic Judaism, theology, spirituality, and the primary substance. He was at war with the inner contradiction, the Old Testament, and the Jewish nation. The Jewish Sabbath, Jewish eating rituals and the Jewish temple and everything else Jewish did not matter to him. He was there to say goodbye. The idea that Yahweh is Jesus Christ is simply mind blowing. If Yahweh-Christ was a conspiracy it was not a Jewish one. Yahweh simply wanted his name back. They pushed each other to the brink and then into the abyss. Passover deteriorated into a slanging match between the God of Ancient Israel and primary substance practitioners also known as the faith family. Gentiles watched the unravelling and seized their opportunity.

Dan Barker was right to suspect conspiracy. The apostle Paul took another message to the Jews of the Diaspora. He restored faith and prophecy and returned everything back to the way it was on Palm Sunday. To seal the deal, he introduced something called *the faith of Jesus Christ*. Being a theologian, he was able to explain to the disciples what they had been through. Peter and James were not theologians. Thirty years after the crucifixion Paul wrote the Book of Romans, a theological masterpiece. He gave the world "amazing grace" and euphoria. Dan Barker used to have a faith healing ministry; he knows the value of euphoria. It is true that we do not have evidence and they did not have evidence that Christ was Yahweh incognito.

What else is true is the intellectual arrogance of the Greek super apostles. They had no need of faith. They found wisdom. They were existentialists. It meant something else to them to hear Christ say, *I and my Father are one.* There was a big reveal in statements like *The Father is in me, and I am in the Father.* There is a name which must not be spoken, *Whoever has seen me has seen the Father.* They could imagine the words coming out of the mouth of Yahweh. But he was not in the habit of explaining his existence. Dawkins enabled Barker and Barker unfettered Christ from the blasphemy laws of his day. The gospels finally elicited a Yahweh. The ex-Christian pulled out the verses from the gospels to make the point that Christ said he was Yahweh. This was supposed to exclude Christ from reality.

In one interview Barker said, "He was violent, and he was angry... a lot of the time. Just like his Father." The ex-Christian could not help but step between Christ and Christianity. He knew the truth well enough to speak it. For sake of clarity Barker added, "He was a chip off the old block. He was not just a chip off the old block, he said he was the old block himself". We could have been listening to one of Yahweh's super apostles. Christ said Abraham was familiar with him and he was the subject of the writings of Moses. He said he was from above and with a past that went back to Abraham. Dawkins was first but now we all see the spaceman. Barker sees the spaceman in the gospels because there is consistency. The Greek super apostles were up close and personal with the spaceman.

Dawkins missed a trick when in the early days he said, *I know you don't believe in an old bearded man sitting on a cloud so let's not waste any more time on that.* After meeting the faith family, he realised he needed all the help he could get. It was time to ask the theologians why they didn't believe in the spaceman. He was hoping they would die of embarrassment. He roped in Stephen Hawking who in 2010 pronounced philosophy (theology) dead. Easy to do if you believe that out of the two it is spaceman who is the God of the Old Testament. Dawkins and Hawking put their reputations on the line for the God of War.

In his book on cosmology Hawking added a powerful statement arguing for the God of War. He said, "*Unless one endows God with some other attributes such as being the God of the Old Testament, employing God as a response to the first question merely substitutes one mystery for another.*" God in sense of creator is no longer an answer to the question, what is the origin of the laws of the universe? Big Bang is the answer. Even if we did not have the answer God is not the answer because God does not mean anything unless the word is defined. Spaceman is a description of the God of the Old Testament. If the word 'God' is defined by theologians, it will expose the primary substance and the history of errors which led to faith and the faith family. If the God of War and the creator God is mentioned in the same sentence, then we are talking contradiction and falsification.

The important thing is that Science made the distinction between the two gods. In 2014 Hawking said, “Before we understand science, it is natural to believe that God created the universe. But now science offers a more convincing explanation. What I meant by ‘we would know the mind of God’ is, we would know everything that God would know, if there were a God, which there isn’t. I’m an atheist.” In 2006 Richard Dawkins revealed Yahweh the God of War. In 2010 Hawking vouched for Dawkins. Science has vouched for the God of War. You could say the God of War being a non-creator idea of God means God has vouched for Hawking and the non-creation of the universe.

# Chapter 9

# Christ As Traction

The Daily Telegraph gave the story front page cover, "Mother in embryo mix-up to give up baby". The story concerned a mother who gave birth to one white and one black twin boys. Would-be parents Donna and Richard Fasano visited Lillian Nash's New York fertility clinic in April 1998. Conception was achieved through IVF procedure (in vitro fertilisation). In vitro is Latin for "in glass". The first test-tube baby was born in 1978 in Oldham General Hospital, Greater Manchester. IVF procedures were unheard of when the nativity stories were written. Matthew and Luke were limited to Immaculate Conception. They could not imagine egg donor surrogacy for a way of explaining the birth of Yahweh.

Immaculate Conception amounted to conception of a kind. Conception was the only method the gospel writers knew to explain the miracle child. It passed the apostolic test because it was not inconsistent with scripture. The story was also one of prophecy. The closer it was to prophecy the closer it was to truth as they understood it. But if we are looking for Yahweh-Christ then there can be no human genes in him because Yahweh is from above and conception is ruled out. The Daily Telegraph told the story of Donna Fasano through facts. Donna was implanted with four of her and Richard's embryos or eggs. However, a mix-up at the clinic caused her to be implanted with several eggs belonging to another couple who had been at the clinic the same day. The other childless couple were Deborah and Robert Rodgers who attended the clinic in the same desperate hope of having a baby.

Matthew and Luke had a hungry public wanting to know more about the man Rome certified a god. The story in broad detail spread like wildfire. His name was on the lips of every Passover pilgrim returning home from Jerusalem. It was simply the most sensational story of their times. He was the talk of the Empire. The details most known about the new god were that he had been rejected, mocked, tried, crucified, certified, buried, and vindicated. Jesus Christ was a name. Whatever people thought about him they had to admire his tenacity. He was distinguished personage. But there was nothing known of his earlier life or his birth.

Both Donna and Deborah undertook IVF treatment comprising of the following basic steps. They were given fertility drugs to stimulate their ovaries to produce as many eggs as possible. The eggs were then surgically removed. In the case of Mrs. Fasano 25 eggs were removed. The harvested eggs were then individually placed into glass laboratory dishes with their respective partner's sperm added for fertilisation to take place. A few days after, some of the eggs, at this stage embryos, were transferred back into the women in the hope of producing one or more healthy babies. A month later there were glad tidings for Mrs. Fasano because she was joyfully pregnant. Mrs. Rodgers again faced heartache.

The clinic wasted no time in notifying the two mothers of the results. But within two weeks of the good news Dr Lillian Nash had to inform Donna of the clinic's mix-up. Unbeknown to the expectant mother the emotional roller coaster of having a baby was about to enter another loop. She took the call which told her "One of the babies you're carrying may not be your child". If you didn't know the birthplace of Jesus of Nazareth, you might look up messiah to find out. You would look to scripture and conclude Bethlehem. You would project messianic prophecy onto the person you consider to be the messiah. *Behold, a virgin shall conceive, and bear a son, and shall call his name Immanuel.* Imagine after all the excitement getting a call from one of the apostles telling you the famous Nazarene and Galilean was not born in Bethlehem Judah as he should have been.

Donna Fasano refused an abortion. Her twin boys were born in December. Donna was birth mother twice and biological mother once. Mary the mother of Yahweh was his birth mother but not his biological mother. If we are talking about Yahweh-Christ, then conception is ruled out. Mary was still scripture's "young woman with child" but via egg donor surrogacy. Yahweh the spaceman does not have a genetic connection with humanity, he is not a product of evolution. The gospel of John tells us repeatedly that Jesus of Nazareth was not from Judea. The Nazarene was rejected out of hand because scripture demanded Bethlehem of the messiah. The Galilean refused to accept it as a disqualification. He blamed his critics for everything that was wrong about him, *Do ye not therefore err, because ye know not the scriptures, neither the power of God?*

The "of Nazareth" becomes a major issue in the gospel of John. *But some said, Shall Christ come out of Galilee? Hath not the scripture said, that Christ cometh of the seed of David, and out of the town of Bethlehem, where David was?* There were people willing to reserve judgement on him, but they were quickly brought back into line. *Are you also from Galilee? Search and look, for no prophet has arisen out of Galilee.* In the first chapter of John there are "origins" verses. Nathaniel said, *Can there any good thing come out of Nazareth?* There is an undercurrent of constant criticism. *Is not this Jesus, the son of Joseph, whose father and mother we know?* We are even told why Yahweh chose to be of Nazareth and it was because *the saying of Esaias the prophet might be fulfilled... Lord, who hath believed our report?*

The gospels of Matthew and Luke tell it differently. They go by Jewish prophecy and Jewish expectation that Son of David will be from Bethlehem. They are reinforcing Jewish scripture. It is as if they wrote for a Jewish audience. Either Matthew and Luke knew for a fact where Jesus was born, or they had no idea. They were men of faith. They were not about the truth. If they had been Russian and all about propaganda, we would understand why they lied. Faith concluded messiah and faith demanded Bethlehem. They were about Jewish supremacy. The gospel was *for the Jew first and also for the Greek.* They told the story as prophecy and not as factual. They could not accept what Jesus said when he said, *David therefore himself calleth him Lord; and whence is he then his son?* He was not the Son of David. They put a faith twist on everything knowing *the just shall live by faith.*

The people wanted prophecy and they got prophecy. Prophecy was consolidated. All the stories of his mundane existence in mundane Nazareth were omitted. Prophecy came from the Old Testament. Prophecy made it *faith to faith.* The apostle Paul started a trend with his spiritual mechanisms. *Christ after the flesh* and *the faith of Jesus Christ* was just the beginning. The nativity stories followed. If the authors knew about Nazareth, then they for the sake of faith chose to be circumspect. John too was written with purpose, *that ye might believe that Jesus is the Christ, the Son of God.* Fact is religion evolved out of the facts faster than the facts could be assimilated. Son of God meant one thing to Yahweh and something else to embryologists. The Greek had to decide for themselves how to integrate the Son of God into their world. Jew and Greek did their best to make sense of it.

Josephus (AD 37 – c AD 100) the Jewish historian commented on what was happening. With a large dollop of sarcasm he said, "the divine prophets had foretold these and ten thousand other wonderful things concerning him". The historian would have known more. He could call upon priests with living memory of the Passover which was thrown into disarray by Jesus of Nazareth. The Jerusalem temple was destroyed in AD70 and his "we told you so" comment to Jewish Christians dates his criticism to post AD 70. His remark is celebratory because the destruction of the temple was pivotal for messianic Jews. Add to that the failure of the 'imminent' Second Coming and it makes sense that the Hebrews started falling away.

Josephus used sarcasm to suggest prophecy had been overused. He managed to accuse Christians of revision without offending any of his Jewish readers. When Yahweh asked, *Whom do men say that I am?* The most common answer was that he was one of the prophets. They believed man is flesh and transmigratory spirit. The spirit made for reincarnation. They did not have IVF in those days, but they have it now. Jesus was not born in Bethlehem. Josephus knew this but when a man says he is God Almighty it is best not to speak of that man. Either Messianic Jews or Greek Christians were liable to feed off his comments.

It turned into a bit of a mess for Donna Fasano. It would eventually be for the courts to straighten things out. Meanwhile, public interest in the story kept growing. Donna had every intention of keeping both babies because they were her babies. Maternal instinct was not conflicted. Certainly, she was birth mother for both. Only two weeks before the story broke the Fasanos learned of the third-party parent. The genetic parents learned of the surrogate mother. The newsworthy story had the public trying to work out the moral implications and deciding what would be the right thing to do. Matthew and Luke released their story of the birth of Christ. They were writing for Jews of the Diaspora and not for the Jews of Galilee. They kept faith with prophecy. They gave it a faith fanfare. But the last thing Baby Jesus wanted was to invite scrutiny.

The Rodgers filed a lawsuit against the clinic demanding full parental charge over any black children born to Mrs. Fasano. Motherhood made headlines in March 1999. The moral quandary was resolved with a sacrifice. The sacrifice was big news. Donna decided to give up one of her boys. It was a sacrifice New York and the rest of the world stopped to admire. Yahweh's IVF procedure was more carefully planned and moral and legal issues were not contested because of the secrecy given it. The story going public would have ruined the project and the leak would have been irreversible.

The IVF objective was to grant the spaceman an experience of life in his own country. They were the children of Abraham, and the people of Moses. They were his connection to planet Earth. He was their geo-political tribal warlord. He had a message to deliver but that would be in due time. They welcomed him into the world. As much as possible it was going to be a level playing field. At thirteen years of age, he was an adult his own person. He was meant to be a person of his times and a person of his place. Yahweh was an existentialist not unlike Navalny. He was staying put no matter the struggle for him to remain an authentic person. The God of War tried to make it a level playing field, but he was disadvantaged by the fact that he was living in a theocracy. Authentic was going to be a painful experience.

Yahweh was back as an Israelite and as a Jew. He needed twelve likeminded people for his mandate, or his message could not be delivered. He needed twelve non-religious Jews because spirituality and the primary substance was his problem. He understood the faith family and anonymity was required. The nice fascists were out to turn everyone into a theologian and that meant special treatment. By remaining incognito he staved off their remedies for apostasy. He was in control if he stayed away from blasphemy. He had to be conscious of what he was saying and in whose presence he was. He could still protest but within the rules. Society set the rules and allowed for non-religious Jews.

Socio-spirituality was thriving and fully vindicated by the natural sciences and Intelligent Design. Yahweh existed only in the third person. People were looking to Bethlehem for the messiah. No one was looking in Nazareth. The rulers of the synagogue demanded spiritual charge over any Israelite child born to Mary. As far as they were concerned, he was theirs. But as far as patriotism was concerned, he was a disappointment. He wanted no part of the Jewish collective. He wanted to change things. John's gospel is correct. The Nazareth fail contributed massively to his reputation. It meant Yahweh could start off from where he left off. Messiah, you must be joking. He was rejected out of hand all over again. It was mutual. They were primary substance practitioners and could not be taken seriously.

"Mother in prophecy mix-up to keep baby". There was no one to help Mary understand the power of IVF let alone the power of God. The handmaiden of the Lord did not know about egg donor surrogacy. But Mary did know she was his birth mother. Mary underwent a unique experience. Donna Fasano had faith in Lillian Nash and her fertility clinic. Her faith was in a promise. She accepted an offer and subjected herself to the procedure which promised her motherhood. Mary had faith in a promise. The promise was in the form of a prophecy found in the Word of God. There was a starting point for her to have an appreciation of the exceptional circumstances.

Mary of Nazareth was kept on a need-to-know basis. Mary didn't know and for a while Donna didn't know she was a third-party surrogate. The author of John's gospel didn't know. He could not write up a nativity story for the Nazarene. But of one thing John was sure, it was not any kind of conception. Jesus Christ came down from heaven. John happily admits Christ was not from Bethlehem. He similarly leaves out messiah and broadens the theme. *The next day John (the Baptist) sees Jesus coming unto him, and saith, Behold the Lamb of God, which taketh away the sin of the world.* John did not have the details of birth. If John had been aware of egg donor surrogacy a nativity story might have attempted. Some looked for prophecy to explain Jesus and some looked to Jesus to explain Jesus.

Donna and Richard Fasano said, "We want what is in the best interests of the child. We are giving him up because we love him." In their own way the faith family has at heart the best interests of every member of the family. The temple rulers had the loyalty of their people. The faith family was of no use to the God of War and again it was mutual. There were many families the world over, but the spaceman expected more of Abraham's family. He said so. He was a social misfit. The closest the spaceman came to being found out was when his wider family said, *Is not this Jesus, the son of Joseph, whose father and mother we know? How is it then that he saith, I came down from heaven?*

They should have called Mary in for a second interview. But they knew he was born in Nazareth and that was all they needed know. When it suited him, he moved the goalposts from messiah to Yahweh the God of War. At Passover the news of Yahweh's visit to his temple was published. For a few months his cover enabled him to rage against the establishment without giving away the fact that he was Yahweh. The temple rulers carried out a thorough investigation into the Baptist's choice of messiah the same week it was announced. They looked for the Bethlehem connection. There wasn't one. It was typical of the prophecy mix-ups they were used to.

The synagogue covering Nazareth knew all about Mary including the young Mary. She had never been away, not even for a weekend break. There was no connection with Bethlehem. In their words, *whose father and mother we know?* Travel was a big deal for some members of the village. The temple rulers left nothing to chance and their file on the Nazarene did not need revisiting. There was no point asking Mary to explain the meaning of *I came down from heaven.* A delegation from the Jerusalem temple had been sent to Nazareth not long into the disturbing messiah campaign. We know they asked Mary of his whereabouts. The primary substance practitioners did not like being demonised by the wannabe messiah. They told him to his face he was the one working for Beelzebub.

There were no virgin birth rumours surrounding Mary. The retrospective nativity stories found in Matthew and Luke set up two very big targets. Real life was different. The target that did exist was the one set up by the Baptist. Matthew and Luke were a part of a conspiracy. Theologians and primary substance practitioners kept shifting the real-life story into their faith universe and faith fantasy. Their theory of everything made sin the big obstruction and the part Yahweh had to play was that of the suffering servant. *The just shall live by faith* was all the excuse they needed for the conversion. They changed the narrative.

Matthew and Luke had to cope with decreased consciousness. They did not know anything about IVF, Evolution, the selfish gene, and Big Bang cosmology. They compensated for their gaps in science with a belief in the existence of demons. They believed in spirituality. They believed in prayer. They believed in theocracy. They believed in social engineering of a faith family. They did not want freedom of speech and freedom of belief. They wanted faith to be given a proper chance to succeed. All they wanted was to feed the delusion. They edited out the God of War and they edited out the war. They ended up with an omnipresent man. They engaged in special pleading, and they conquered sin. They were euphoric about it.

Before spiritualisation by the gospel writers, Yahweh had months of fun screaming and shouting at theologians without them knowing why. Religious Jews were given a taste of their own medicine. They could not understand him, where he was coming from and what he was about. He was introducing anti-faith ideas thick and fast. He frustrated them with parables as they frustrated him with literary devices. They chose the primary substance over him for God. They chose Saul over him for king. Galilee saw the inner contradiction came to life and battle it out for supremacy. Skygod vs theism was the hottest ticket in town. He spent months irritating them till they had to shut him down all over again. They knew all about his God of War but were having none of it.

Meanwhile, his Galilean followers thought they were seeing messiah being defined. Every time he tried to establish the office and authority of messiah he was cut down. Only he knew what was within his remit. Angry was what it meant to be messiah. God never prayed to himself, and they were forced to ask him to pray. He talked about Yahweh in the third person because at the suspicion of blasphemy they were ready to stop their walk with Jesus of Nazareth. Still, they were some of the happiest days of Yahweh's life because there was no one comparing him with the primary substance. They were all too distracted by the thought that he was the messiah of prophecy. It was their job to get him to Jerusalem safely.

The Manhattan fertility clinic got it wrong. They could not fix the maternal problems of their making for either Donna or Deborah. Donna had one child with her genes and one child genetically not hers. Both babies were her flesh and blood. The vital mother services which provided growth and development came from Donna. Birth defined the mother. Deborah was the biological mother and her and Robert's genes meant they had a prior claim on one baby. Or so they thought. It was beyond the clinic to find the answer. The bond between mother and baby only got stronger. The clinic failed because they lost control of their own processes. The clinic lost all control when the case was passed to the courts. Brand new moral issues made it of public interest. Motherhood was put under scrutiny.

It was new territory for the courts and justice needed to be worked out. Donna told her story, "*I don't want to deprive my son of being with his biological mother. Any woman who has given birth to a baby, cuddled, kissed and fed their baby, will understand how I feel today. Who cares what colour my little boy is? I carried him, gave birth to him and loved him. But he needs to know who his biological parents are and will want to be with them as he grows up. And I understand how this woman must feel, wanting her boy, because I know what it is like to be a childless couple.*" Dr Lillian Nash commented, "We've tried to do the right thing all along. I've never had anything happen like this before".

Things changed for Donna Fasano when she signed up for IVF treatment. She was fully committed to the process. The clinic's procedures succeeded in delivering two babies. The process had a flaw which led to problems she could not resolve. The parents resorted to legal action. Going to court helped because they were all willing to abide by the court's decision. The court was given control. It was emotional turmoil for both sets of parents. Donna's boys bonded with each other. The right thing for her to do was to separate them. She was compelled to speak out. "She may have given me two beautiful babies, but she destroyed their lives. To them, it may be a job – to me it's my life. This wasn't my doing. People with infertility problems should be able to go to their doctors and trust them to do the right thing".

You could say Dr Lillian Nash stretched the truth and stretched motherhood when she took Donna Fasano to third party surrogate. You could say messiah to God of War is stretching an idea beyond recognition. Those who believed in prophecy were left behind when messiah suddenly was the God of War. It explained Jesus of Nazareth for some people and made him untouchable for others. It was easy for the spaceman to say to theologians, *You err, not knowing the Scriptures, nor the power of God.* Scripture blocked his path. Scripture demanded Bethlehem and Elijah and the Son of David. The anomaly put himself out of reach of scripture and prophecy.

Remote conception and embryo transplants were unheard of in Galilee before the 20th Century. Primary substance practitioners had no idea that human embryos could be frozen till it happened. The Word of God did not give that kind of advantage. The first "frozen embryo" baby was born in 1984. She had been frozen for 8 weeks before being successfully unfrozen and implanted back into her mother's womb. Doctors had perfected the freezing technique. When we combine embryo transfer and freezing the story of the spaceman gets more interesting. We can attempt to answer the old question, *how is it then that he says, I came down from heaven?* Conception does not help any when you are trying to understand, *Ye are from beneath; I am from above.*

IVF research was started in the 1970s by Patrick Steptoe and Robert Edwards to treat female infertility. It became existential or a reality in 1978. In 2023 the science continues to bounce off churches because the Bible is the infallible Word of God. Christians are sold on the Immaculate Conception. They do not want to admit error because error leads to deception and deception leads to delusion. There is only one way the spaceman can be accommodated and that is through change. There is so much they did not know 2,000 years ago. There is so much more Christians of today know but they insist on a system which deceives itself.

One question defined Jesus of Nazareth. *How is it then that he says, I came down from heaven?* There was a change in status from skygod to earth-God. Frozen human embryo is a game changer. Frozen means suspended. An embryo suspended in Bethlehem could be unsuspended in Nazareth. The period between suspended and unsuspended does not have to be in weeks. It could be years. Never mind "stretch" prophecy because it turns out God can toy with it. God found a way around prophecy and the temple rulers were none the wiser because they lacked the science of embryology. It says a lot about what was happening on the ground that someone should ask, *how is it then that he says, I came down from heaven?* Meaning, we need to incorporate heaven into our fertility theory.

For someone to come from "above" today the IVF procedure must take place in space and that could mean the international space station. For our purposes, the test tube baby will be a space baby or from above. For that same embryo to come from Bethlehem the procedure to implant the egg into a host womb must take place in Bethlehem. If nine months later a baby is born in Nazareth, then we can say the child is from above and Bethlehem and Nazareth. If a Bethlehem connection can be established, we are back on course with Old Testament scripture. Christ came from above, Bethlehem and Nazareth.

The temple rulers asked a key question, *how is it then that he says, I came down from heaven?* They knew what they needed to know to ask that question. He believed in Yahweh the spaceman. He believed he was Yahweh the spaceman. He said something to make it clear he was the spaceman who had come down to Earth. His Greek super apostles believed it to be the case. What did they know? The temple rulers wondered what they had missed. How could it be the case that he came down from heaven when everyone knew he was the son of a Nazarene couple. According to the gospel of John he said, *I am the bread that came down from heaven.* It was the *came down* which prompted their question. But he did not fall out of the sky and so in what way did Yahweh come down intact from heaven.

Back to human fertility, baby girls are born with about two million eggs. But whilst in the womb, the foetus can have as much as seven million. By the time puberty is reached it is more like three hundred thousand. It sounds a bit science fiction but if born with all her eggs this means the eggs inside a woman were once inside her mother as she herself was. The temple rulers were satisfied Mary could not be placed in Bethlehem in and around the time of his birth. If not Mary, then her mother might have visited Bethlehem. She might have been there to ask Rachel for her blessing for the child she was carrying. If an embryo transplant took place in Bethlehem and it was suspended for years like a frozen embryo, then Yahweh could feel satisfied with himself.

Thanks to Richard Dawkins we know enough about the crazy side of Yahweh to know that he would. The God of War was willing to go to extraordinary lengths. The embryo transfer/egg donor surrogacy procedure had to take place in Bethlehem for the spaceman to enjoy torturing his critics over prophecy. It was more fun for him if it turned out that he was from Bethlehem. If an embryo can be suspended for years, then the wait for the special child started from before Mary herself was born. The temple rulers were thorough in their investigations. The validity of Jesus of Nazareth was checked and double checked but there was no connection to Bethlehem, none that could be found.

The temple rulers lived by rule of law. Scripture gave them no choice but to reject him. If you are set up to be the child of prophecy you must live up to prophecies. It was the Baptist who started the prophecy trip. They indulged prophecy as far as possible. Yahweh is not the messiah and stopped pretending in the hope that they would see him for who he is. They failed to recognise him. The God of Ancient Israel wanted his rejection replayed. The story of unrequited love began with the Golden Calf. The story continued into Wednesday of that

Passover. The God of Abraham, the God of Isaac and the God of Jacob became the God of the whosoever. God's experiment in existence or his experience of life was predominantly an existential crisis.

Amanda Holden was four years old when she became an existentialist. She wanted to be taken seriously. According to Amanda her biological father abandoned his children. We are interested in her side of the story, and it is what she says in her autobiography. She was ignored when she was desperate for her existence to be acknowledged. The message coming through was rejection. A vivid memory was created. The experience left an emotional scar. Decreased consciousness was forced upon her by the absence of a parent.

The exemplary existentialist reacted to being casually dismissed. Amanda Holden from then on wanted to be taken seriously. She needs to compensate for the esteem which was withheld from her.

The four-year old is still in her and is always seeking to compensate for that moment of unimportance and rebuff. Amanda wants to be recognised and she wants as much recognition as she can get. Ambition plays a big part in that quest. Yahweh was not being taken seriously when the children of Abraham employed the golden calf. He never got over that day or the day the prophet Samuel told him Israel did not want him for king. Instances of idolatry told him his people did not want him for a god. Then came philosophy and Intelligent Design and theism and spirituality. Academia ganged up on him. The spaceman was rubbed out of Jewish consciousness. He became a figure of fun.

He was desperate to be taken seriously but any attempt to reinstate the spaceman into Jewish consciousness was laughed out of court. To this day he is greeted with howls of laughter by theologians. God too suffers emotional scars and physical. He needs to compensate for the esteem withheld from him.

He wants to be recognised and valued. He wants as much recognition as he can get. Ambition plays a big part in that quest. His existential challenge was befitting a god. He picked his moment. It was a golden age for the Jews when he decided to visit the temple in Jerusalem. History repeated itself because no one was interested in talk of the spaceman.

It was time to test his mettle on a level playing field. A level playing field was not so difficult to achieve. The spaceman only had to come down from above. Once again it was a case of what you see is what you get. His go at nation building had led to theocracy. Both parties knew all about the prophets. The Word of God provided common ground where they could meet and argue it out. His people welcomed him into the world and into their world. When the time approached for him to deliver his message from God, he adorned the mantle of messiah. He was given a platform and high visibility. He kept the spotlight on himself with miracles of healing.

He had the pleasure of personally reprimanding his priests. They had no idea where he was coming from. Revenge was served cold. He put himself through the experience because he needed to prove himself to himself. He needed to find himself in the tumult. He is an existentialist and he was following through as a man of his times. Navalny is an existentialist and he stayed in Russia on the same impulse. He too has a message to deliver to the people of Russia and they should take him seriously. He too is surrounded by theologians or their agents. Christ proved his resilience and resolve was equal to theirs.

In the Garden of Gethsemane, he reached an important marker in his life. He learned to let go. He submitted himself to due process. He gave up control to procedure. He hit the headlines with Yahweh-Christ King of the Jews. The world got a look-in. God was at one point of his existence small as a baby. At another point small as an embryo. He was at one point no bigger than a selfish gene.

## Chapter 10

# Another Theory of Everything

The Second World War was year zero for Europe. The supremacist agenda ended in disaster even for supremacists. In October of 1945 Jean-Paul Sartre captured the new spirit of the new age with a new creed called existentialism. Supremacy was somewhat blunted by the fact that everyone is existential. New was added to by Fred Hoyle who coined the phrase Big Bang. The long goodbye started in 1949. Big Bang was coined by someone who did not believe in it or agree with it. Hoyle was arguing for a static universe as opposed to one which is expanding. The radio interview triggered widespread discussion of Big Bang. In the 1960s the discovery of cosmic background radiation was evidence for Big Bang. The world said goodbye to the primary substance which was a substantial part of its old self.

We need a brand-new universe to go with the latest idea that God is a man. The Holy Trinity is a spiritual universe which is a derivative of the primary substance. Faith is the operating system for all spiritual universes. In the spiritual universe God is a man only if the man is also a philosophical entity.

Meaning, God is something of a shapeshifter. Or it might be that God is a two dimensional being. Jesus is said to be all man and all God. In one dimension he is all man. In the other he is an omnipresent philosophical entity which happens to explain how creation came about. The Holy Trinity is an attractive idea of universe, and it is loved by millions of theologian Christians. But the Holy Trinity turns the God of War into something less than a shapeshifter.

We decided to take the spaceman seriously. For us the spaceman exists. Christianity is a problem because the spaceman is reduced to a literary device. We end up losing God to anthropomorphism. The reality of God is eroded. The shared history we have with God is eroded. Spirituality changes the narrative, and the story is given a faith agenda. We seem to be stuck to the primary substance. But there is a universe where it is possible for God to be a cosmological creature. To get there we just need to tweak Big Bang. Common sense says a man could not have created the universe, but we are not looking for a universe which has been created. When it comes to universe, we find ourselves on the side of Stephen Hawking and Spontaneous Creation. On the Socrates-Aristotle barometer we lean toward the Unmoved Mover and incidental creation.

Einstein said, "*I don't try to imagine a personal God*". Closer to the truth it was not possible for him to imagine a personal God. It was not possible for either Hawking or Aristotle. It is not possible for Richard Dawkins. A personal God is something you do have to imagine because God is not a personal God. I cannot have a personal relationship with Elvis because he is dead. I cannot have a personal relationship with Yahweh-Christ because he has left the building. Anyway, it is a universe we are looking for. It is Catholic priest Georges Lemaître (1894 -1966) we are looking for. The Big Bang theory has many fathers including a philosopher-priest turned scientist-priest. Lemaître like everyone else was deeply affected by war. He served as an artillery officer in the Belgian army during the First World War. Quite sensibly, he did not want another war.

Lemaître published a paper as early as 1927. It was about *"A homogeneous universe of constant mass and increasing radius..."* The Catholic priest was a science giant. He was first to propose the universe is expanding. By the way, constant mass means fixed measure of mass and therefore finite. The mass is not increasing, and it is indestructible. The temptation is to say infinity but that is an attribute which our universe does not require. His paper accounted *"for the radial velocity of extra-galactic nebulae."* But before anyone could catch up with Lemaître he worked out an origins story for the universe. He reversed his expanding universe to as far back as he could take it. The shrinking universe took him to a Primeval Atom. He imagined a cosmic egg exploding the moment of creation.

The universe was given an alternative beginning to that given it by philosophers of old and copycat theologians. There was no way of attributing the new science and universe to either Moses or St. Paul. The science of Big Bang begins with Einstein's Theory of General Relativity published in 1905. Excellent science but we are more interested in what happened afterwards. We are interested in the theory as it was understood by the Catholic priest. Increased consciousness came by way of reading Einstein's paper. It was Einstein's cosmic constant at issue. One gets the feeling the priest was looking for change. World war can do that to a man. When we look toward Russia, we know Christianity is a dangerous thing. The priest was doing his best to keep Europe from another Christian war.

The theologian said, *Once you realize that the Bible does not purport to be a textbook of science, the old controversy between religion and science vanishes.* The old controversy between religion and science is the theory of Evolution. It is the biological soul as opposed to an immortal soul. It is the existence of the Christian God and its improbability. It is the whole scheme of spirituality and divine attributes which have no basis in science. Lemaître lived in a world before Richard Dawkins and his God of War. Dawkins might be tempted to say that because of its God of War the Bible is not even a textbook of theology. Before everything else there was a hot dense primordial soup of matter. The unanswered question of Big Bang is what caused the hot dense primordial soup of matter.

Religion and science have been polarised for a long time and not just since the publication of *The God Delusion*. Subatomic particles are the building blocks of the universe. Particles turned into atoms. Atoms to molecules to gases before we get to Genesis which picks up the creation story at a later point. There is a substantial amount of cosmological detail left out by creationists. Theists misappropriate science to justify a readymade universe. But science evolved to a point beyond the readymade world of Genesis. Dawkins went back to the Bible only to find there was no science to refute. It was based on philosophy. The science of Anaximander and Socrates was flawed. What they thought of as natural we think of as supernatural. Spirit was part of their natural world.

Before we understand science, it is natural to believe that God created the universe. Dawkins took a turn at describing the natural sciences as it was known to St. Paul. *"He not only created the universe; he is a personal God dwelling within it."* Very introverted and very strange. Lemaître the scientist-priest professed the Bible is not a science book. This is huge. For someone who lived through times of great tension between Christian countries the openness was more a cry for change than literary criticism. It was time for the church to let go. Science and religion have a shared history but in today's world it seems like one is holding the other to ransom, one is holding the world to ransom. It seemed that way to the world's most intelligent man of 1927.

It was his great love of theology which prompted him to call for a separation of religion and science. Dawkins the evolutionary biologist ended up meeting Bible scholars and many of them did not think Science was their concern. They were experts in spirituality, and their job was to serve their communities with the most profound idea of God available to man. Science was a secondary item and evolution as a part of science was seen pretty much as a non-event. Their excitement came from the two-way relationship with God which was thought to be possible. As far as they were concerned it was a possibility. In the days of philosophy God used to be a matter of science. Dawkins wanted to expose flaws in their science but what he discovered was the absence of science.

Religion and science already separated sometime between 1927 and 2006. Separation occurred naturally and without any fanfare. The $20^{th}$ Century had developed into a fork with two prongs. In 2006 Dawkins published "The God Delusion" and became a household name. He spent what seemed like a few years promoting his book. It made him a media personality. His many appearances were always entertaining, and this was helped by the fact that theologians did not understand evolution. The professor's patience was tested many times and he started to develop an edge. But it was promoting and campaigning overlapping because he was also a champion of militant atheism.

The great God debate of 2006 ended with scientist and theologian polarised. But what we learned was that it went back to at least 1927. Religion paid lip service to science giving the impression the disciplines were reconciled. The Catholic Church first expressed enthusiasm for the Big Bang theory in 1951 claiming it was scientific validation for a belief in the existence of God. But God is sometimes a loaded term. Lemaître knew he had challenged the method of creation with a view to forcing separation. It was more like creation and the expanding universe was superimposed on top of the omnipresent God. Therefore, philosophers made a mistake about creation but managed to stumble onto theology. As far as he was concerned *the old controversy between religion and science* came to an end in 1927. Big Bang was a fresh start for both science and religion.

The existence of the universe makes it impossible for me to believe in absolutely nothing. I exist and you exist. The fact is things exist. Better still, matter cannot be destroyed. The method of creation from emanation to Big Bang has increased the age of the universe from thousands to billions of years. The Big Bang model of the universe is not the model of the universe the Bible has. Readymade is out of the question. We need a model of the universe where God can be a man and not a primary substance. Big Bang takes us in the right direction. Ten billion years ago there was nothing out here. The universe had not reached this far out in space. Twenty billion years ago the universe did not exist. There was a time without time. There was no creator to speak of. There was no God of War and spaceman.

Socrates and Plato paid no regard to Yahweh the God of War. They were not even aware of his existence. In turn, they were of no interest to him. A child of the universe is not a child of God. The children of Abraham were the children of Abraham and not the children of God. The God of War said to them, *I will take you to me for a people.* Moses famously reminded him they were his people. Yahweh-Christ kept reminding the Jews they were the children of Abraham and a disappointment. They had a different take on life or as the God of War said *Search the scriptures; for in them ye think ye have eternal life: and they are they which testify of me.* Primary substance practitioners thought a mind meld with the primary substance was possible and that meld was a taste of eternal life.

I know of at least one person who has stopped thanking God and started thanking the universe. As part of the universe, she has a kind of a relationship with her universe. It is not a personal relationship because the universe is not a person. Theists have made a person out of it but the "person" they have manufactured is not worth knowing. Faith is required to access that person because that person is not existential. 1927 was a good year. 1929 was an even better year. Astronomer Edwin Hubble provided evidence that the universe is expanding. He also determined there are other galaxies beyond the Milky Way. The universe is made up of island galaxies. According to the Son of Man those other galaxies are desolate. According to the primary substance they are not.

A Catholic priest proposed a method of creation without relying on metaphysics. Einstein caught up with him in 1935 and conceded his understanding of the Theory of General Relativity was correct. The new method freed the metaphysical God from design and maintenance of the physical universe. God was still omnipresent, but the universe was not. God was infinite, but the universe was finite. God was fixed but the universe was expanding. God and universe are no longer attached. God is spirit but man is matter. Yet, primary substance practitioners were able to keep their working system working. Somehow God the metaphysical and man the physical still achieved a mind meld. In the new universe introverted and meditative does not necessarily get through to God.

The further back in time we go the smaller the universe. Smaller means increase in density and heat. One million years ago the galaxies were more densely situated. One million years into the future and the galaxies will be even further apart than they are today. The size of God mattered to the philosophers, but it is one justification too many. Do we really need the attribute of omnipresence? It makes no difference to the man in the street if size is limited to one galaxy or one solar system or one planet. I am sure other people find size just as annoying as I do. It must be the most annoying thing about spirituality. It is up there with demons. Light years of Jesus Christ is not awe inspiring. It is an insult to my intelligence.

Religion and science managed to go about their business without treading on each other's toes. After all, they are about different things. The idea of a readymade universe lost traction. It was the turn of the primary substance to be driven into the margins. Faith was in decline, but the faith family decided to put up a fight. They hijacked technology and science and used it against the world. Militant atheists decided to fight back with their own weapon of mass destruction. They too could fight dirty. They hijacked the Bible and the God of War and used it against the faith family. They made the connection between the spaceman and Jesus of Nazareth. Faith was given an existential crisis. This war has created an uncomfortable living space.

It is one thing to know about God and something else to have a personal experience or a personal relationship with God. Aristotle and Lemaître made the distinction. We can know about God in the same way we can know about Jesus of Nazareth. That knowledge is plenty, and that knowledge is inspiring and the basis of a personal commitment. We can be emotionally affected. We can respond and we can react to the crucifixion of Christ. That is as personal as it gets. We loathe the God of War, and we can fall in love with the character of Christ. Lemaître was different to Aristotle in that he was still advocating a two-way relationship with God and the need for theology. But his method of creation meant it was time to let go of science. Faith for faith's sake and spirituality for spirituality's sake was the way of the future.

When a scientist says something can come out of nothing and the universe is entirely of itself then "nothing" becomes a loaded term. We end up endowing nothing with potential and attributes. If history repeats itself, we will end up with a nothing that has personality and ought to be worshipped. Creator is a loaded term because creator means an omnipresent philosophical entity which has feelings and fatherly characteristics vouched for by Hellenistic Judaism via the Old Testament prophets. It is so easy to start corrupting language and applying specialist meaning to everyday words for purposes of social control. Christianity is full of double talk because Jesus of history is also a philosophical entity. If the hot dense primordial soup of matter is as far back as science can go then regardless of what triggered Big Bang it remains proof that absolutely nothing does not exist.

The hot dense primordial soup of matter might have poured into our dimension from a completely other dimension. Are we the nowhere the soup boiled over into? Before we start going off in all sorts of directions, we must remind ourselves that we are looking for a universe to go with the spaceman. We are looking for a universe which can support the God of War. We are not trying to prove God is the creator. If anything, we are trying to prove that God did not create the universe. Sub-atomic particles are the building blocks of the universe, and they go back to the hot dense primordial soup of matter. Particles make everything today 13.7 billion years old. Big Bang did come after Anaximander's primary substance, but it is a universe without the primary substance.

Our God is the God of War, and we know two important things about him. He was for a while as small as a single selfish gene. The other extraordinary fact we have about him amounts to an origins story. He was begotten. When the angry selfish genetic God was forced out of the House of God, he said to the primary substance practitioners *your house is left to you desolate.* He was taking God with him. Aristotle's primary substance was ultra-pure and ultra-holy. Christ was flesh and blood and material. He was not even spirit let alone ultra-pure and ultra-holy. He was not spiritual. He told the primary substance practitioners of Galilee that the spirit of the Lord was upon him. That qualification meant God was with him wherever he went. He was a living Psalm 139. He told his Greek super apostles he was begotten.

Is there anything pure God about the man Yahweh-Christ? God is not flesh and blood. An atom is smaller than a selfish gene. A particle is even smaller than an atom. How much of the spirit of the Lord would it take for Yahweh-Christ to be God? One particle of God is all that is needed for God to be God. Only one particle would need to be begotten of God to be the God particle. When we say begotten, we mean literally and in the original sense the word was spoken. St. Paul introduced theology between twenty to thirty years later. Even if it was one day after the return to above the sky it was after fact. Everything about him is after the fact. In the universe which is for the God of War the divine particle was begotten out of one dimension and into another.

Begotten is the process by which the God particle became a part of the hot dense primordial soup of matter. When Christ said begotten, he was not talking cosmology. He needed to explain how it was that he was the God of War. The Greek apostles knew it was the spirit of the Lord which made him deity. That part of him which was divine was begotten of God. Richard Dawkins found the God of the Old Testament, but you do not have to be a genius. Jesus of Nazareth found the God of War. He persuaded some he was that God of War. They were given no more evidence than Dawkins. The difference was his apostles did not think it was silliness. They believed the God of War to be the true concept of God which it is. We can take the principle of begotten and apply it to the creation of the universe.

Big Bang is our universe and not the primary substance. Begotten means begotten and not eternally begotten. Science takes us back to the hot dense primordial soup of matter. Yahweh-Christ takes us back to the God particle which is all the God the universe has. Empty space is good for our theory because it would make begotten emphatic. A very small point punctured an empty dimension to fill it with the hot soup of matter. There are a few things we can deduce about the God particle. It is physical as much as any other particle is physical. It is the oldest and smallest thing in the universe. It is like all matter, indestructible; and it was described by Jesus of Nazareth as the spirit of the Lord. The God part in him was something like one part in ten trillion trillion.

The story of God is about begotten. It is a story about two dimensions. On one side is *Particle Number One* and matter and on the other is the substance which begot *Particle Number One*. The soup followed hot on the heels of the special particle. The hot dense primordial soup of matter is made up of *Particle Number One* and the energy it took to create it, or the energy released by its creation if you prefer. The energy could not be contained. The divine particle ended up away from the source and nowhere. Begotten was the method and *Particle Number One* was a deliberate creation. What followed the hot dense primordial soup of matter was Spontaneous Creation. It was incidental creation and God acquired the status of Unmoved Mover. Begotten and not Intelligent Design was God's involvement in creation.

*Once you realize that the Bible does not purport to be a textbook of science, the old controversy between religion and science vanishes.* It vanishes if we let go of the idea that God is the creator. It is then we can *endow God with some other attribute such as being the God of the Old Testament.* There is no pretence that Jesus of Nazareth or the God of War or *Particle Number One* created the universe. There was a time when Jesus of Nazareth did not exist. There was a time when the Earth did not exist. There was a time when the Milky Way did not exist.

There was no sign of a God of War in the hot dense primordial soup of matter. There was a time before the hot dense primordial soup of matter and a time without *Particle Number One.*

God did not create the universe and he is not *a personal God dwelling within it.* In one universe science is sacrificed for faith and in the other it is faith which is sacrificed for science. Without faith it is not possible to have a personal relationship with the primary substance. Without the primary substance it is not possible to have a relationship with the primary substance. It is agonizing to lose a friend but what is needed is to redefine the terms of that relationship on basis that God is a man. Even then we cannot have a two-way and close relationship because the man is off world. The faith system is a working system, and it leads to euphoria. It is possible to continue it. But it also leads to world war. It contributes to war.

Christian Russia is a country without freedom of speech. Truth is warped on two counts. A supremacist agenda passes for patriotism. The men behind the propaganda have generated a siege mentality. It was once true, and Russia was under siege. They won the Second World War, but the war cabinet wanted to stay in power for ever. They kept the siege mentality going. Her people endured repeated bouts of starvation into the 1990s because it was patriotic to starve for the motherland. They love patriotism. Some hardcore supremacists acquired Christianity; it did not change them. They remained self-serving whilst dishing out faith to the poor. It made it easier to get away with excesses. Oligarchs are supreme for no other reason than being Russian. God help anyone who comes between oligarchs and patriotism.

Christian Russia is a theocracy in which it is a crime to offend religious feelings. Truth is warped on a second count. A supremacist streak runs through spirituality and demands totalitarianism in the form of theocracy. The men behind the theocracy have generated a persecution complex. Socialism had the resources to win the war against impoverished flesh, an unfair world, and a cold-blooded devil. The religious leaders wanted to stay in power for ever. They kept the persecution complex going. The people endured a test of faith lasting centuries. They are willing to sacrifice truth, freedom, and life to be a good Christian. God help anyone who comes between them and their faith in God.

Particle Number One started life as a stand-alone particle but by the time of Abraham it was part of a quantum mechanical device to be known as Yahweh. It was as Yahweh that the real work of acquiring a personality began. Philosophers attributed eternal personality like they attributed everything else to the eternal and immutable primary substance. Meanwhile, Yahweh the God of War started the journey of life in a customised gene. God is not a man although you might mistake him for a man if you did not know any better. If we dare to take the spaceman seriously, we must at some point consider his genetic make-up. The Yahweh gene is the best God could come up with. *Particle Number One* was to assimilate the Yahweh gene and it had to be fit for purpose. It led to the ethnic cleanser. Assimilation was not meant to be easy. The gene had to assimilate *Particle Number One.*

The gene gave him social instinct. The spaceman is a social being. The journey of self-discovery required socializing. First contact was with Abraham who was a huge influence on the new and evolving personality. For a while the aspiring friend of Abraham was known as the God of Abraham. God was a bit obsessed with Abraham. His search for identity started with a tribal lord. It continued with Isaac and Jacob. Moses was last to play a huge part in the evolution of Yahweh. *Particle Number One* and the gene achieved perfect assimilation in the Garden of Gethsemane. When Jesus said he was God it was only half the story. When he said he was a man it was only half the story.

*The God of Abraham, the God of Isaac and the God of Jacob* was quite literally a personal God for Abraham, Isaac, and Jacob. The rest of the world's population did not feature in his comings and goings. It suited him to start with individuals. He was new to emotions. He was volatile. He had a lot of growing up to do. The selfish gene made God about acquisition. He was about acquisition of personality and identity. Richard Dawkins wrote up a personality profile to go with the selfish gene. *Particle Number One* on its own does not generate a personality profile. From the point of view of humans, a dot is no existence at all. If God exists, he must wonder what it is like to really exist and have feelings and be a person.

Reverend Monsignor Georges Lemaître, a Belgian scientist, challenged the scientific establishment of the day with his Big Bang theory. He held out an olive branch when he said, *the Bible does not purport to be a textbook of science.* He introduced science which was alien to the Bible. It was a view of creation St. Paul had not anticipated. The important thing was that the controversy between science and religion could end. The controversy was more like a war. Militant atheists and Christian fundamentalists existed in 1927 and they were at each other's throats. It was unseemly. The scientist-priest wanted to stop the fighting or at least soften attitudes. It was correct for him to say *the Bible does not purport to be a textbook of science.* But it was not something the Catholic church was ready to say.

Lemaître was ready to say science *does not purport to be a textbook of* theology. He did his best to introduce his science to the church. Thing is the theology of St. Paul relies on the science of St. Paul. It was good old-fashioned religion based on good old-fashioned science. Lemaître strengthened the case for evolution and the biological soul. Nothing undermines spirituality more than the biological soul. The old controversy between religion and science was made worse by the new controversy. The controversy cannot vanish without religion and science coming together and speaking as one. Each is needed to validate the other to some degree. In 2010 the conflict worsened when science assumed theological expertise and declared the God of the Old Testament was a spaceman.

The quantum mechanical upgrade to a biological creature meant consciousness for *Particle Number One.* Before then the God particle was a divine portal for God the remote source and nothing more. At last God could mean something more than a dot in the world of man. When first contact was made with Abraham the particle in a majorly way became aware of itself. If we want the God of War, we must admit there was a becoming. Yahweh is not a readymade God. His consciousness was tingling with his becoming a social being. He had to take things slow and one friend on Earth was more than enough for him to be dealing with. Meanwhile, God could carry on interfacing with God.

The portal and interface did not stop being a portal and interface. It just became more complex. The nature of the information changed. It was all about the biological upgrade and how it was assimilating creation. Yahweh-Christ could say about their unique relationship. *Whoever has seen me has seen the Father.* But in the Garden of Gethsemane assimilation was completed. *Particle Number One* and customised gene was finally his own person. *O my Father, if it be possible, let this cup pass from me: nevertheless not as I will, but as thou wilt.* The first example of Yahweh acting as his own person was when filled with jealousy, he demanded Abraham sacrifice his son. God proper intervened and stopped the proceedings in time to stop the dagger piercing Isaac.

The original Galilean disciples were described as unlearned men. The original Greek disciples were described as chief apostles or super apostles and they were some of the best educated people in the world. They observed and made a study of Jesus of Nazareth. They concluded he was the God of War.

They understood how God worked and the interface relationship. They did not think God the Father sent God the Son to save humanity from sin. It is Yahweh's story, and he takes the lead. The only rule is righteousness. It was righteousness which put him on the cross. If we analyse the story, we find it was the remote source who was aware of Judas's actions. Judas betrayed Yahweh and that was unforgiveable. But he was willing to destroy his own reputation as a price for stopping Plan A. That was righteousness God could not ignore.

Yahweh-Christ had mind set on revenge and vindication. Three times he asked for Plan A. He was acting like his own person. The righteous God had to do the righteous thing and the most righteous course of action for him was to submit to due process. Plan A was aborted, and plan B was instigated with the Lord's Supper. His was a personal battle to achieve recognition for Yahweh the God of War. In that respect Plan A and Plan B were the same. They were to the same end. Theology did not come into it.

# Chapter 11

# Qeiyafa Shrines

In 2012 the discovery of artefacts provided physical proof of a belief in Yahweh. Israeli archaeologist Yosef Garfinkel presented to the press two model shrines found at an excavation site southwest of Jerusalem. The ruins at Khirbet Qeiyafa were first noticed in 2003. Digging began in 2007. The excavation uncovered a city eight acres in area with two monumental gates and a wall running 770 yards in circumference. The first dramatic find from the site was a ceramic shard which contained the oldest example of Hebrew ever found. The writing supported the case for the Bible's authenticity because it meant 3,000 years ago the Israelites could record the stories which made their way into the Old Testament.

Carbon dating of burned olive pits dated the city to between 1020 and 980 BC which is the time of King David. Archaeologists found bones of sheep, goats, and cattle but none of pigs. It suggested the inhabitants followed a dietary prohibition on swine. Animal bones enabled Garfinkel to make the important qualification that the city was inhabited by early Israelites. Garfinkel said, "*Over the years, thousands of animal bones were found, including sheep, goats, and cattle, but no pigs... Now we uncovered three cultic rooms, with various cultic paraphernalia, but not even one human or animal figurine was found. This suggests the population observed two biblical bans, on pork and on graven images, and thus practiced a different cult than that of the Canaanites or the Philistines."*

Creationists laid siege to the find claiming it a part of their story. The Iron Age fortress city was theirs because the absence of figurines constituted evidence of a belief in an invisible immaterial transcendent omnipresent creator God which was a forerunner of the primary substance. But not so much a philosophical entity as a spiritual one. Archaeology concluded, "*Perhaps the most notable aspect of the Qeiyafa finds is not what has been found but what has not.*" However, the model shrines found in 2012 are equally significant if the connection we should be looking for is with a spaceman. We know the wanderer god of the Ancient Israelites was away from heaven for the years he spent with them in the wilderness.

The absence of figurines is explained by the fact that they were used to having him around. It was like there was nothing better for a god to be doing. He was someone with time on his hands. Before his nation building phase the God of Abraham was a god without portfolio. He was the god who was a slow starter. The gods of Egypt and Canaan had temples. Yahweh was a god without a temple and the Israelites were a people without a temple. Both God and his people were challenged with nation building. Shrines are mostly for the worship of a distant heavenly creature. Figurines represented skygods. Not so with Yahweh who was making Earth his home and Israel his business. In the early days his people suspected he was the same distance away as when Moses was his project manager.

The Ancient Israelites had good reason for keeping the shrine empty. The heavenly creature was not without representation and was maintaining some sort of presence through judges. He was not then an absentee landlord. He was around the proverbial corner keeping an eye on his investment. The spaceman had made his intentions known. Yahweh the tribal lord was excited about nationhood. The Israelites shared his enthusiasm. The God of War was a big part of their consciousness. A long-distance relationship did follow because they asked for a king. They wanted one of their own for king. The request wounded Yahweh. He assumed he was their king. Their relationship needed to be worked out. Meanwhile, he obliged them.

The prophets made a study of Yahweh through his relationship with Moses. The slaughter which followed the Golden Calf incident was a major lesson. He could be provoked. They built up a personality profile. It was their job to maintain a fear of the Lord. Communication flowed in both directions. The people were able to request a king. Moses helped maintain a safe distance between the volatile God of War and the people of Abraham. The go-between was required to represent both sides. Yahweh's reactive nature or bad temper and their failure to adapt made their relationship impossible and stalled his plans of integration. The people did not want to be anywhere near him on a bad day. But he had invested too much in his project to completely give up on it. He was stuck with them.

It is true the starter religion of the early Israelites did not have figurines to represent the skygod come down to earth, but it was nothing to do with theology, philosophy, creationism, and spirituality. Their Yahweh came down from heaven of his own free will. No one sent him into the world. The friend of Abraham was become a tribal lord by his own doing. He was ambitious, he wanted geo-political presence in the world, and he needed Abraham's stock for that. The opportunity to escape slavery had the Israelites following Moses into the wilderness. The ban on graven images reflected the intimacy they had. They were one in purpose and for that they needed to look to their history and not the sky.

The exodus generation did not have figurines of Yahweh, nor did they have shrines. But what they did have was Yahweh in person. They had the real thing. Their God had his feet on the ground and was never far from their camp. They were not allowed to invade his space but at the same time they were assured of his presence by an accompanying pillar of fire by night and a pillar of cloud by day. Moses gave him daily reports on how things were going in the camp. Yahweh was familiar with some of the names although never having met with any of them face to face. Moses asked to see God's face but was denied. He was there to do a job and not to make a name for himself. He was there to write the first draft of Scripture. The scriptures were his testimony of Yahweh the God of War.

If the empty shrine of Qeiyafa said anything it was that the spaceman from above was not above when the shrine was in use. He was never far away from the people of Moses. It was a different world before Anaximander. The ban was nothing to do with the Primary Substance or the Supreme Being or the Unmoved Mover or any other philosophical entity. It was a time before Anaximander's creator agent which permeates creation. It was a world in which it was OK to say Yahweh. The cult of Yahweh was different to that of Canaanites and Philistines, but it does not mean that it was that different. Skygods had different histories. Different does not mean the God of the Israelites was galactic, inter-galactic, multi-galactic and all-pervasive.

With skygods the sky is the divide. With Yahweh the divide was the prophet. The people wanting Saul for king was a slap in the face for both the grand one and his prophet. It was a humiliation. It was one to put it down to experience. The God of War was left in no doubt that he was going to need a plan B, C, D and possibly E. There was bound to be another "next big thing" around the corner. The journey he had started was going to be full of setbacks and frustration. He took the hint and stepped back and out of the picture. He became an absentee landlord. He was pushed back to his side of the sky. It was going to be a challenge just to remain a part of their consciousness. The next time he was going to face his people they would be strangers to each other.

The discovery of shrines without figurines makes for a strong connection with Moses the lawgiver. *And the LORD (Yahweh) said unto Moses, Thus thou shalt say unto the children of Israel, Ye have seen that I have talked with you from heaven. Ye shall not make with me gods of silver, neither shall ye make unto you gods of gold.* For the Moses generation the spaceman came out of the past as much as he did out of the sky. It was the God of Jacob. The probability is that Jacob wrestled with a spaceman when he said *I have seen God face to face, and my life is preserved.* We of a modern mind are embarrassed by the story but that was not the case in the time of Jacob.

Seeing the face of God was a milestone in the life of Jacob. The baton had truly passed from Isaac to him, so to speak. *The God of Abraham* and *the God of Isaac* was from that moment *the God of Jacob.* The rest of the world was excluded from a personal relationship with the spaceman. The children of Jacob were familiar with the story. The children did not think of themselves as the children of God because they were the children of Israel. Jacob lived before Anaximander and before Moses and there is no reason to suspect he thought of God as the creator. He thought of God strictly as *the God of Abraham* and *the God of Isaac.* He did not have another way of defining God. The only other thing he knew about God was the colour of his eyes.

Yahweh made it clear to the Caiaphas generation that it was Abraham who was their father and not God. God was his father and not theirs. God was his father because *Particle Number One* was begotten of God. Abraham was not like his children. He had better manners. Abraham was receptive and allowed

Yahweh to have his say. He was courteous. The theologians were in the habit of covering their ears and shutting him down. The Moses generation knew it was Abraham's God who had sent him. They knew this as soon as they learned of his name. Memory of God and Abraham survived. The covenant of Circumcision helped keep those memories alive. Yahweh meant something to them. *I am that I am* applied to him in a very real way. His name was a clue.

The spaceman was a work in progress. What you saw was what you got. The stories of Abraham, Isaac and Jacob defined him, and the stories were well known to the Exodus generation. If the story had stopped at the point of God giving his name, there would be no God of War. There would be no meaningful list of unpleasant characteristics. His personality was still evolving when he said, *I am that I am.* Nation building was the next big step in his journey of self-discovery and for that he needed the people of Abraham as much as they needed him to get out of Egypt. The exodus generation had the privilege of knowing his name. The baton had truly passed from Jacob to them, so to speak. Theirs was a fresh venture. It meant they could stop calling him *the God of Abraham, the God of Isaac, and the God of Jacob.* He was their God.

The commandment not to make figurines was made in realtime. Figurines were not required because Yahweh had entered their lives in a dramatic, spectacular, and memorable way. Figurines were not needed to achieve consciousness of Yahweh. Their story was his story. Revision changed the narrative but not well enough to get it past Dawkins who ended the conspiracy of theologians. When figurines did show up, they were of other gods. Figurines showed up in a big way under the rule of King Solomon. They provided validation for deities of his many foreign wives. Figurines were banned and locked away during the reign of Hezekiah. In 2012 the discovery of artefacts provided evidence of a buried past.

# Chapter 12

# The Pilate Stone

Probability for the existence of Yahweh went through the roof in 1961 when evidence was found for Pontius Pilate the co-founder of Christianity. In their time Pilate and Christ were the two most hated men in Jerusalem. They spent one Passover outdoing each other. Theirs was a mutual admiration society by the time they met face to face. The idea of having the priests of Judea beholden to him was not without appeal to the Roman. Seeing them panicked was the source of great pleasure. It was a desperate situation for the priests because the one known as Christ was growing in popularity with the same Gentiles they had invited to Passover. The Christ was a rival, and he was turning the Gentiles against the hosts. The Temple hoped to reverse the trend.

Having the blood of Christ on their hands would have been a scandal too far. Galilee's anomaly was become Jerusalem's curiosity. Pilate's own wife was getting sleepless nights because of him. A good Roman would not normally be attracted to a foreign God, but no other God was offering what Yahweh was offering. Guarantee of afterlife, eternal wellbeing, citizenship of heaven and a mansion thrown. It was all hers if she could make the controversial decision to add him to her collection of gods. The wife of the Governor could not take sides in the Jewish quarrel but that was only protocol. It was opportunity others were not letting pass. She was torn. Pilate was face to face with the man who promised so much.

It took one eventful week for Galilee's outcast to become distinguished personage. Christ had powerful enemies, but he also had powerful friends in his Greek super apostles. Both the temple rulers and the super apostles knew all about imperial protocol and procedures. The Jewish administration had first claim on Jewish citizens. A Jewish citizen like a Roman citizen was guaranteed certain rights. Rome's Governor was caught in the middle and sensed more jeopardy than opportunity. He refused to take the case. Rome and Jerusalem argued over whose jurisdiction applied. The temple rulers were demanding urgent recourse to Roman law. They dragged him in chains to the foot of the Roman Governor and dumped him there. Christ the controversial figure was guilty of one controversy too many and the Jewish administration was washing its hands of him.

Christ was free to negotiate with Rome directly. It would eventually be for Rome to grant him a kingship anyway. It was their empire. But the narrative had changed overnight. His kingdom was no longer of this world. He was not there for a kingdom. There was nothing the Roman could do about a case which was outside of his remit. The situation was made more difficult by the fact that justice had just been dispensed. The Roman court was in effect double dealing. Pilate's priority was to avoid recrimination. It was something which could cost him dearly. The focus from beginning to end was on procedure and procedure directed the narrative. Circumstances brought the two protagonists face to face. But by that time the God of War had turned his back on war.

More than correct there is a lawful narrative which comes with the Roman magistrate and Roman law. There is a legal statement made through the court. There is jurisprudence and legal precedence. Anyone found changing the narrative is guilty of perverting the course of justice. Theologians think they are exempt because they are justified by faith. Anything they do for the sake of faith is justified. They have no problem distorting history. They say without faith it is impossible to please God but by God they mean the primary substance. The original narrative belonged to Pilate the Roman Governor of Judea and the overseeing authority. The priests took the case to court, and it was for them to prove the defendant was a liar, a fraud, and a trickster. The job of the court was to make it a fair trial.

There is a certain narrative which comes with Yahweh the God of War. Richard Dawkins found that narrative and used it to produce a personality profile. Not a single theologian was able to say he was imagining the God of War and that one did not exist. Yahweh's super apostles did not need to be geniuses to work out there is a God of War lurking in the Old Testament. Once you know the Old Testament is about the God of War and you know something of his personality then you are ready to read the gospels. Dan Barker recognised Yahweh in Christ. The super apostles were not disadvantaged by the gospels. They quickly ruled out prophet and messiah. Anomaly was the word for Jesus of Nazareth. They worked out he was God because they knew the Old Testament as well as anyone.

The God of War has been eliminated from Christian consciousness. Every Christian the scientist met was a theologian or a victim of the faith propaganda. Dawkins was at a loss. Dan Barker took possession of Dawkins's Yahweh. The Bible expert recognised the conflict between the Old and New Testaments. The first part was about a God of War and the second part about the primary substance. Pilate and all Gentiles visiting Jerusalem did not have to answer to the New Testament. They were aware of gods of war like Mars and Ares. Gods of war are about winning at all costs. No one could deny the God of War of Ancient Israel had put up a valiant fight against some of the best theologians in the world.

There is a certain narrative which comes with philosophy and theology. It is a form of special pleading to see it in terms of salvation and rescue from sin. The father of spirituality said, *All things arise from a primary substance and return to it, unless they pay fine and penalty for their iniquity*. He was wrong on both counts. Judas handed over control of the narrative to the temple rulers with his betrayal of Yahweh. Yahweh handed over control of the narrative to the temple rulers by submitting himself to due process. He further yielded to Rome and the Roman magistrate found himself in control of the narrative. St. Paul came like a thief in the night and stole the narrative off everyone else. He proved to be a master manipulator and he saved Hellenistic Judaism from the God of War. The supremacist said the gospel is for the Jew first.

To their credit the temple rulers stopped to reconsider the merits of Yahweh the God of War. But Christ failed to impress them. Whoever you are, not everyone is going to like you. The men of Galilee lost the narrative when they found Isaiah 53. Paul's letter to the Corinthians confirms the idea that Jesus died for sins was taught to him. It came from the bona fide apostles. The unlearned disciples were heavily influenced by John the Baptist. Prophecy seeped back into the narrative. The anomaly was once again Messiah. They were all over the place because Christ had made it clear he was not the son of David but the Lord of David. It was God's guilty pleasure that Judas had carved out a path to greater righteousness. The betrayal was the best thing that ever happened to him. For an existentialist it was the apex of existence to be able to say I exist.

Mission accomplished, victory sealed, vindicated and finally atop the emotional plateau befitting his lifelong struggle. It was a personal best. *Particle Number One* and the customised gene were fully integrated in a way which was unimaginable

24 hours earlier. Between the Lord's Supper and the Last Supper there was the Garden of Gethsemane incident. From an existentialist's perspective personal fulfilment was worth every cut and bruise. Procedure played its part. Eleven out of twelve disciples played their part. The super apostles played their part. The temple rulers played their part. Scripture played its part. The people of Jerusalem played their part. Rome did not want to play her part, but Pilate was pressured and for the most part he played his part. Judas, the God of War, and Pilate did not play their parts.

The Roman historian Tacitus said of the politico-religious association "checked for a while, broke out again". We know Christianity grew and grew and kept growing till it became the official religion of the Roman Empire. Through many narratives it started growing because it was all the returning Passover pilgrims could talk about. It was a news item and a headline story. Pilgrims returned to Rome, and it was the talk of the town. Only the super apostles understood it in terms of a new religion. St. Paul needed at least ten years to work out a gospel for his missionary journey. He had his theology all worked out when it came to writing the Book of Romans. At first the story spread and kept spreading by itself. It was fascinating to think a skygod had come down to Earth. The only time Christianity stalled or was checked was the day of the crucifixion. That day was pivotal.

In 1961 evidence supporting the historicity of Pilate was unearthed. The archaeological artefact known as the Pilate Stone was discovered in Israel and is so called because on the block of stone was an inscription bearing the name of the opportunistic servant of Rome. The inscription like the one on the cross suggested the Roman officer was adept at making friends and influencing people, the right people. It was not untypical for a middle ranking officer to keep favour with his superiors. Tributes and dedications fell into the right category of gesture when wanting to ingratiate oneself with those higher up the chain of command. The dedication on the Pilate Stone was for Emperor Tiberius. But we know Pilate for another one of his tributes.

Rome's magistrate managed to write himself into the history books with his other inscription which said, "Yahweh-Christ the King of the Jews". In 1961 he again came to the rescue of his spent friend. Pilate is a constant reminder of the God of War who stood face to face with him. The two protagonists came together at that most famous of trials which ended in the death of Jesus of Nazareth. Yahweh the warrior warlord proved he could fight clever but maybe a little too clever for some people. He relied on Pilate to play the part of High Priest. He made the declaration. The outcome from their unholy encounter was that in death he was granted a precious operator's licence. The temple rulers had refused to endorse him and even ejected him from their temple. The natural sciences or at least the theoretical physics part of it convinced them of a primary substance. It meant the man had to be a fraud.

There was not a threat to Rome because his kingdom was off world. It was as much off world as the kingdoms of Mars and Ares. It was on the other side of the sky. The difference was his kingdom was open to the people below who at that critical time were siding with him. He was not coming back for them because they would be going to him. Pilate acclaimed him a God on a par with Mars and Ares. It was easy for him to do so because he was not a theologian, and he had no interest in a personal relationship with an intrusive invisible friend. As far as he was concerned divine philosophy was divine delusion. The magistrate interrogated Roman law for any possible legal justification for the Jewish goings-on. The temple rulers explained the involvement they had planned for him. They would support his decision. But nothing was going to save him if he failed in his duties.

The Jewish citizen was subject to Rome's laws the moment the Jewish court took away his Jewish citizenship. He was physically transferred to Rome within a matter of hours. In fact, as soon as they were open for business. He was not given time to break any of Rome's laws. It tested the legal officer to find a transgression in the time it took to transfer him. Rome would not normally interfere in religious matters. That would be for the local rulers. Jerusalem was a theocracy, and every aspect of life came under scrutiny of the primary substance and was policed by priests. Their Pax Romana specifically forbade the Roman governor interfering in Jewish affairs. He was invited to interfere. Bad call on part of the theologians.

The Temple's plan to involve Pilate backfired. He was not subject to Jewish law. The Governor of Judea was the only one in Judea who could get away with saying Yahweh. Not even Yahweh was allowed to say Yahweh because when he did, he found himself guilty of blasphemy. Not that it was something a God of War would be worried about. The magistrate did not want to say Yahweh and cause a major international incident. Yahweh himself spent a lifetime engaged in subterfuge. The town was on edge as it was. But there was a job to be done and at some point, in capacity of magistrate he would have to say Yahweh. It would be at that point that control of his court and the narrative would pass back to him.

The office of messiah which had been touted in Galilee was of no interest to Pilate. The messiah campaign ended on Tuesday of that Passover when he relinquished the office of messiah with the statement that he was the Lord of David and not his son. Rejection of everything he stood for was made emphatic by his being ejected from the temple. That was when he could not say Yahweh, but Pilate was not a Jew. Pilate could talk openly about what was going on. He could say Yahweh without breaching the commandment *thou shalt not take the name of the LORD thy God in vain.* He would be respectful whenever speaking the name. Dawkins never said Yahweh in vain because every time he said the name it was in context of it being the name of the God of War. Yahweh was said with intent of causing a crisis of belief. The Roman found himself between two parties accusing each other of mistakes, lies, conspiracy, deception, and delusion.

It was problematic for Pilate to be dealing with the case. He was there to defend Rome's reputation and interests and not to re-evaluate the Bible's inner contradiction. His office was being abused. Judgement had already been passed by the Sanhedrin court. The prisoner was brought to him bruised and chained. If the punishment did not reflect the severity of the crime, they needed to reconsider. They had proved he was not invincible. He was a beaten man. There was still the option of prison and exile. But the faith family wanted more than justice. They wanted vindication. They wanted to close the book on the God of War. They wanted a return of law and order. But their idea of law was the law of faith. The faith family dispensed faith from inside a Roman court.

It was chaos. Being the God of War was against the laws of faith. Pilate worshipped sky-gods and from his point of view it was the misfortune of one to touchdown in a theocracy. The magistrate was desperate to find a crime befitting the punishment already under way. The punishment which had been set for sedition. Pilate was looking for anything which resembled sedition. His political career depended on it. He needed to take the initiative and fight clever. He found a crime but to pin it on Jesus of Nazareth he needed to be guilty as charged. The presence of Yahweh the God of War would be a challenge to the Roman order. The Romans had their cult of the emperor and Tiberius ruled by divine mandate. The divine mandate bestowed on the emperor divine status.

Whoever was emperor could act like he was God. He could command people to bow to his image. Roman supremacy and empire were based on the divine mandate. The Greek super apostles believed Yahweh the God of War had come down from above. It was like Jupiter or Mars visiting Rome. Yahweh did not have divine status bestowed upon him because he was already divine. His was a superior mandate. The Governor defended his emperor from the Jewish usurper. History bore out the threat anticipated by the astute servant of Rome. Pilate was a Roman soldier, and we would be surprised if Mars did not head his collection of deities. The temple rulers were primary substance practitioners, but Pilate was not, and neither was Jesus of Nazareth.

Hellenistic Jews named the primary substance Yahweh. Paul an ex-employee of the Jerusalem temple was working under guidance of the same primary substance. The one he had known all his life. He was a religious Jew. He converted to Christ taking his primary substance with him. He was not willing to give up the primary substance and ended up doing what the scribes of old did, he gave it a new name. The same happened with Gnostic philosophers whose primary substance was called Logos. They renamed it Jesus. The most important thing was to get the name right. That's how we got the multi-galactic Jesus who is bigger than a London double-decker bus. A Jesus who is anything but. For theologians a God of War does not make sense. It's not like you can have a personal relationship with a God of War. You're better off staying away from them.

Christ's super apostles were mostly from Jerusalem's Greek synagogue. They knew all about the primary substance because they used to be practitioners. They used to search the scriptures for eternal life. Mind meld justified their practices. But then they happened upon the temperamental Yahweh. Intellectual arrogance was the key to their success. They believed in their own comprehension skills in the same way as Dawkins and Barker do. It was possible for them to know a lot about God but to call it a personal relationship was a bit of a stretch. They gave up their phoney friend for the anomaly they suspected was the God of the Old Testament. Jesus of Nazareth was leading the way against error, lies, deception, delusion and coverup. Skygods were back in the ascendency and Pilate could appreciate that.

Pilate did not know if the Galilean was the God of War, but he knew the super apostles thought it was the case. They were willing to fight and die for him. They were on hand to explain the story of the God of Israel to fellow Gentiles. In the week Yahweh took back his name he first announced his arrival to the Gentiles of Jerusalem. Word spread to Gentiles of the Roman world. It started with approximately five hundred anti-theist Greeks. St. Paul in his letter to the Corinthians addressed them as the enemy. He accused them of corrupting his Jewish gospel. In seeking to discredit them he ended up providing his readers with a ton of information on the super apostles. We know their leader was a man called Apollos.

Back to the trial where the people who saw Jesus that day did not for one second think, they were looking at a multi-galactic philosophical entity taking time off from sustaining the universe. They hated him because he refused to go along with their delusion. The God delusion had become a national pastime. They had created a theocracy around the delusion. They forced the God of War into the margins and then off the page. The God of War came down with the alarming message that he was going to purge his land of theologians. God was just as much capable of doing a little rejecting of his own. He gave them a personal relationship with God, but it was not the kind of personal relationship they were looking for. They were his personal enemies, and their time was up. But on the night that he was betrayed he was compelled to change his mind. His rage against them stopped. Their rage against him boiled over.

The kingdom was no more of this world which meant Pilate could not get caught up in the moment and destroy the theocracy to safeguard the future of mankind. He was left sober in the cauldron of passions. The kingdom was no more of this world and there was nothing he could do to help rid the world of the delusion and the deluded. The God of War had ruled out an alliance. He had run his course and was at the end of it. Civilization would have to run its own course and face annihilation at the hands of another theocracy. (Not even Georges Lemaître could prevent the inevitable moral decline of the Christians.) There was nothing spiritual about either Pilate or Apollos or for that matter God. By the time the inscription was put on the cross it was clear to everyone in Jerusalem that God of War was the issue.

Pilate had plenty to work with for his court report. The gospel writers had plenty more to work with. But they did not have access to Christ, Pilate, or the High Priest. When Mark wrote his gospel it was for a check on all the strange things he was hearing during Paul's missionary journey. He broke away to write his gospel. He had dozens of witness statements to work through. His gospel reveals how little they understood of key themes. Gentiles hardly get a mention. Son of God is translated into primary substance. Sin became a driver and prophecy drove them to Isaiah 53. Gnostic philosophers followed suit and translated Son of God into Logos. That led to a gospel where believing Jesus is the Logos was its purpose.

The gospels are unkind to the co-founder of Christianity. But Pilate was used to dealing with arrogant and conceited theologians. The ones he faced were desperate to control the narrative. They wanted Christ consigned to oblivion. They wanted the truth consigned to oblivion. Meanwhile, Gentiles were fascinated by and sympathetic to the tribal warlord who had no choice but to turn against his own people. Midway he changed his tactics and decided to ignore them. He was all about the whosoever which invariably meant the Gentile. When he refused to defend himself that just added mystery to the drama. It was theatre, it was a tragedy being played out. He talked the talk but rather than walk the talk he stopped talking altogether. Those who reserved judgment on him were left waiting for a show of power. They were not given evidence of a God of War.

The prize was the narrative. Pilate put the all-important inscription on the cross while Christ was alive. Pilate handed the God of War a strategic victory They stole the narrative from the temple rulers. Yahweh's existential crisis was over. It encouraged further mocking from primary substance practitioners because it seemed to be a hollow victory. Very soon the title would mean nothing to him because he would be dead. There was no point sticking around and feeding the righteous indignation of the crowds. When Pilate heard the man was dead, he could not believe it. There was no reason for him to be dead so soon. He had it checked. He gave permission for the body to be taken down. The God of War did not die for the sins of the world. He died to get his name back.

In 1961 we discovered the Pilate Stone. In 2006 we discovered there was a God of War. In 2016 the dots leading to Yahweh-Christ were connected. Yahweh and Pilate co-authored the inscription found on the cross. It redressed a great wrong. It made up for the six thousand times the name Yahweh was taken out of the Old Testament. They gave scripture new context. Theologians introduced spirituality and assigned authorship to the prophets. Theologians introduced spirituality to the Yahweh-Christ story. Again, they made sweeping changes. They perverted the course of justice. They made the trial a waste of time. They showed the same reluctance in calling Yahweh by his name. They were about their own thing and their thing was faith. The narrative was changed in order not to contradict faith.

Skygod believers were sympathetic to the plight of Yahweh. He was the skygod most badly affected by the spirituality and faith which proceeded forth out of the primary substance. Gentiles were present in Jerusalem that time of year Jews showed off the benefits of spirituality. The skygod was there also. Gentiles were used to collecting and swapping gods. There were so many gods to choose from. Yahweh the skygod was an interesting choice. We learn from Tacitus the historian that sympathy for the God of War which flooded Rome started in Jerusalem. Skygod believers could imagine their own skygod doing a few rounds with theologians. Yahweh the skygod was fighting for his life. He had to be admired for the way he went about it. He managed to win the argument.

Pilate certified Yahweh-Christ. Yahweh who was a part of Jewish history became just as much a part of Roman history. Yahweh-Christ was not a philosophical deduction or a divine revelation. He was that nuisance of a man who was relentless in his pursuit of finding someone who would take him seriously. It was all about the narrative and Yahweh did his bit and Pilate did his bit. Yahweh had more reason than any other skygod to come down to Earth. He had been rubbed out of consciousness by theologians. He wanted to restore consciousness of himself. A unique set of circumstances led to Pilate declaring the man on the cross *YAHWEH THE KING OF THE JEWS.* The Governor was asked to change his tribute to something not so damaging to faith. The magistrate stood firm. He said, *What I have written I have written.*

# Chapter 13

# The Greek Finally Get Existentialism

Lucian of Samosata (c. AD 120-180) was a Second Century novelist and essayist. He was a celebrity author of his times. He started as an educator. After becoming proficient in the works of Homer and Plato he took to the road with his brand of education and entertainment. His career in public speaking enabled him to travel. He was good at observing human behaviour and giving it a fun twist. Any public figure was fair game for his wit and satire. Lucian could exaggerate and mock and pin it on famous people. He was entertaining, informative, and popular. He was not lacking in imagination because he is credited with being the first science fiction writer. He was driven. He wanted to understand people and their world. He was well known on the circuit.

Lucian had his own inimitable style. He knew how to grab people's attention and that was by starting with a bit of gossip or news before giving it his treatment. We are dealing with someone who was professional and deliberate in the things he said and the way he said them. But his commentary on life tended to be funnier still for being true. He was a man for the truth. He worked hard to get to the truth. We get a good example of the man in his essay "*The Death of Peregrine*". It is about a famous death of a famous man of those times. Peregrine the Cynic chose the Olympics of AD 165 for leaping into fire and burning to death. Lucian blamed Christians for the death of his friend.

The essay is famous for what it says about the Christians of the day. His criticism is in the form of a letter to a friend. It is not a work of theology. In it he laments the death of a colourful character who was known for his brushes with the law. A Cynic is Socratic with respect to philosophical bent. In style of his mentor, Peregrine's death was a publicity stunt. He ended his life throwing himself into fire but not before publicizing the event to the crowds attending that year's Olympics. He left hundreds if not thousands with a haunting memory. He made a spiritual statement. He put dualism into the news. He left for posterity his afterlife affirming testimony.

(Dualism was described by Richard Dawkins, "The dualist believes the mind is some kind of disembodied spirit that inhabits the body and therefore conceivably could leave the body and exist somewhere else". Peregrine went with the science of his day. The science of today suggests innate dualism is a better description and a more likely explanation of the psychological phenomenon. Peregrine would have known about Aristotle and his biological soul but that ruled out afterlife. A dualist believes afterlife is attainable. It is a gift of God to those who have faith. Faith is the operating system for the spirit universe. Christians did not believe in spirit or afterlife because Christ said, *If a man keep my saying, he shall never see death.*)

Lucian was convinced his friend's venture into Christianity did not help him any. Peregrine did not believe Jesus died for his sins and did his own dying. He was someone who converted to Christianity and then went back to being a Cynic. Peregrine the Cynic was not your typical convert to Christianity. According to Lucian the philosopher was a modern-day Socrates. He was an academic and he knew his stuff. The Socratic believed God the Creator played an active part in daily life. People needed to let God play his part. Being a Cynic was letting God play his part. The theologian believed faith brought God into a person's life. God was as good as real. The Cynic tried out Christianity. At some point the intellectual started to toy with Christians just because he could.

The Cynic Christian decided to do his brothers a favour. He would turn them into Cynics without them even knowing about it. Such was their ignorance and his superiority over them. According to Lucian the Cynic had an almighty impact on the Christians of Turkey. You could say he brought Christianity to an end. Emperor Trajan made it possible for Christians to become a part of Roman society. His policy of pardon through repentance meant it was in their interest to go back to their old lives and continue contributing to the temple economy. For a while they lived a double life. Christianity survived and Christianity thrived. By the time of Peregrine, they were become ambitious and competing with Cynics for souls.

Greek Christianity was all the Christianity there was in AD 111 when Pliny wrote up his report. That was because Jewish Christianity stopped expanding when Paul's missionary work came to an end. Jewish Christianity went into sharp decline after the sacking of Rome in AD 70. Messiah had nowhere to return to. Jewish Christianity died when prophecy failed, and the imminent Second Coming of Jesus Christ did not take place. There was nothing Jewish about the Christianity of the Second Century. Christ was not the Son of David because he was the Lord of David, and he was from above. There was not a single Jewish gene in him. Lucian of Samosata was Greek by ethnicity, and he was passionate about all things Greek.

It must have come as a shock to Lucian the discovery that Greek Christianity was Greek culture because it came out of Corinth and from the mind of a Greek genius. It was the religion of Apollos the super-apostle. It was more Apollos than Jesus in the same way Jewish Christianity was more Paul than Jesus. Christ trusted the narrative to his apostles and of them some were known as super apostles. This means the Christians who were the subject of Pliny's letter to Trajan were of Apollos's Greek church. They were being written about again 60 years later. Thanks to Pliny we know something about the Christians who were despised by Lucian and Peregrine.

Lucian was an expert on the works of Plato. He knew all about Greek philosophy. He read many books. He read many books on metaphysics not unlike the book of Romans because every Gnostic leader had a plan to save the world. Christ himself could be interpreted a Gnostic leader and a modern Socrates of his time. Any philosopher interested in making his mark on society could see the advantage of assimilating Christ. A Christ figure could be made to champion the theology of another Apollos or St. Paul. This was already happening in Rome. Gnostic philosophers had a name for their philosopher-king. Jesus of Nazareth set the standard no one could match. He was the measure of Christ.

Peregrine was a modern Socrates, and he could play philosophy with the Christian sect. Lucian the investigative reporter was true to his calling. He researched his story and little details give that away. Big details give it away too. We learn so much about the Christianity of his day. The standout comment must be "*The Christians, you know, worship a man to this day*". We have a connection with Pliny's Christians because they sang a hymn to Christ as to a god. When the modern Socrates converted, he did not require an induction course. It was a religion for simple souls and Lucian even called them misguided creatures. It was still big news that a skygod had come down to

Earth. It was big news that he wanted to share existence with the whosoever. Christians were about celebrating Christ.

Lucian was a reliable intellectual who went further into Christianity than Pliny the magistrate. He went back to the beginning. Not only did he admit the historical Christ he could tell his readers some of the things Christ said. He could believe in Christ but not in Yahweh-Christ. There was a kind of evidence for Yahweh, but it was circumstantial and anecdotal. It was all talk. He said of Christians, *All this they take without evidence.* There was no evidence that the man Jesus Christ was Yahweh the God of War. All the evidence was lost in the Garden of Gethsemane. Lucian was careful not to get drawn into discussions about the Jewish God of War because that only assisted fantasists. Lucian spoke of Yahweh-Christ only as Christ.

Lucian was Greek and Apollos the Corinthian was Greek. Lucian had a good education, what we would call a college education. He was a success in his chosen field. He was an academic and an intellectual. He wrote himself into this book. Apollos was a genius and someone whom success could not evade. He was Greek aristocracy. It was because of him the Greek apostles were known as the chief apostles or super apostles. They understood the God of War and what he was trying to achieve. They went to his emotional rescue at his time of need. From Wednesday of that Passover, they were the ones to be found by his side. They were there for his personal protection. Lucian was fascinated not by God but by the Greek super apostle.

Apollos did not share in Lucian's admiration of Homer and Plato. His admiration was for God. Apollos thought he was God's friend because he accepted the testimony of his friend who said he was all the God there was in the Jerusalem temple, Judea, and the world. Apollos the existentialist was a mystery and that bothered the cultural powerhouse. Apollos who could have achieved so much in the natural sciences and academia chose to give himself to the work of building a church. The theologians of the First Century managed to survive the God of War. The only evidence of his visit to Earth in the Second Century was the Christian. In other words, those who still worshipped a man. Lucian was obsessed with Apollos and somewhat frustrated by his failure to understand the architect of Christianity.

Lucian had to know why someone with such a brilliant mind behaved so erratically. Where did he go wrong? Something led him astray to come up with the ridiculous idea that God is a man. Apollos must have known what he was doing and that meant the failure to understand him was on the part of Lucian. Truth did not often elude the writer, and he needed answers. His Greek identity was important enough for him to travel to Corinth in search of the truth. The investigator was famous for his love of truth and all things Greek. He was welcomed by the Christians of Corinth. They were happy to talk to him. He was someone who could help disseminate the gospel. He was given access to church documents. He read documents in Apollos's own handwriting.

Lucian became very familiar with Apollos. His curiosity was satisfied. It was another famous Greek to add to his Greek consciousness. We learned from Pliny that in AD 111 Greek Christianity could be traced back to AD 86. Apollos who wanted to fight was of fighting age at the time of the crucifixion. He was a super apostle which means he was super before acquiring the status of apostle. It is not unreasonable to think he lived to a ripe old age. It is not unreasonable to think that by the end of the First Century he was living memory. The church in Corinth was visited by Paul. But the church existed long before he embarked on his first missionary journey in the 40s. The Greek Christians were ahead of Paul and had their own Greek apostle in Apollos.

Lucian absorbed everything about his Greek hero. He was qualified to speak on Apollos and the church he fashioned. He knew what marked them out from their beginning. *The Christians, you know, worship a man to this day*. He wanted to know why they worshipped a man. He wanted to know why Apollos thought such a thing. Apollos knew all about theology but rejected it. He rejected Greek philosophy and went back to the worship of a skygod. Lucian knew that Apollos knew all about Socrates, Plato, and Aristotle. He knew about Anaximander. It was a logical step to go from creation to a creator in principle. Intelligent Design supported belief in a creator. Lucian was mystified.

If Lucian was going to understand Apollos, he would have to take the subject of his worship seriously. He described Jesus of Nazareth as *the distinguished personage who introduced their novel rites and was crucified on that account*. He was sympathetic. Jesus of Nazareth was a man who had to be admired. He changed the Roman world with introduction of a new social class. Although academics and theologians thought it daft to worship a man it was not necessarily Jesus's fault if simple folk took that position. Jesus the Christ created a happy family who had to sing about it. Society was friendlier and more open for all members of the collective. They enjoyed their intimacy. It was a new social experience which many enjoyed. There are a lot of people who go to church for the social life and there is nothing wrong with that.

Pliny recorded how Christians worshipped the man who was *distinguished personage.* To stop them worshipping he made political association unlawful. He had to stop them assembling. Pliny found Roman citizens and slaves in the same community. They stood side by side singing their hymns. It was an idealistic and revolutionary concept of society. It was their idea of paradise. Pliny tortured two female slaves to get out of them a more sinister motive they might have been hiding.

They were not trying to deceive anyone. By the time Lucian was writing about them society was no longer troubled by their quirky habit of assembling before dawn. The slaves assembling 60 years later were doing the same. They enjoyed the euphoria which came from celebrating togetherness together.

Servants were not the best educated people in the Roman Empire. They received only the education needed to carry out their duties. It was easy for Lucian to be arrogant and dismiss them as gullible. They enjoyed the singing, but did they know what they were singing about? Naturally they liked the new idea of a society where all men and women are equal before God. But there was a rich history to their religion. What did they know about Christ? Apollos the super apostle was a superhero for having carved out beautiful ideas out of Judaism. The Greek had to fight Jewish Christians to keep them from stealing his Christ and his religion. Lucian the researcher knew more about their story than they did. There was a lot more substance to their morning ritual.

Christianity was a religion that was not a religion because it was a religion without a God. They worshipped a man. Absurdity was keeping out intellectuals. The system worked hard to keep theologians and philosophers out. The fact is Lucian's research into Apollos made him an authority on Christianity. He knew Christ did not die for the remission of sins. He died for another reason, "*who introduced their novel rites, and was crucified on that account*". Lucian knew what those novel rites were. They met before dawn to celebrate the Lord's Supper. *It was their custom to depart and to assemble again to partake of food.* The original group broke company after the Last Supper to assemble again. The betrayal and Garden of Gethsemane incident instigated change. The apostles needed to regroup. The Lord's Supper celebrated the political association which made for the Greek church.

Political association was another way of saying social revolution. Lucian managed to get hold of intimate details. *It was impressed on them by their original lawgiver... that they deny the gods of Greece and worship the crucified sage.* For a lot of people, the crucified sage can only mean the resurrected Christ. The instruction came after he was risen. *Distinguished personage, original lawgiver, and the crucified sage* strongly suggests Lucian was an admirer. It was just the lazy Christians he objected to. The death of his dear friend Peregrine affected him badly and he looked for someone to blame. He was convinced someone was to blame. Meaning what he knew about Christianity led him to believe his friend had been let down. If only the Christians had known more about their own history, it might have helped. He had a dig at the Christians.

*You see, these misguided creatures start with the general conviction that they are immortal for all time, which explains the contempt of death and voluntary self-devotion which are so common among them.* They were not misguided. The idea that people could escape death went back to Apollos. Lucian was one of those who would have said to Christ, *Now we know that thou hast a devil. Abraham is dead, and the prophets; and thou sayest, If a man keep my saying, he shall never taste of death.* It is very simple. When Apollos died his body died but at the same time his body did not die. His biological soul did not perish. It benefitted from a certain quality you find in matter. He was indestructible. Meanwhile, primary substance practitioners and sympathizers believed spirit is substance. It was like water, and it had properties.

Lucian found another problem with the gullible Christians. *They despise all worldly goods alike, regarding them merely as common property. Now an adroit, unscrupulous fellow, who has seen the world, has only to get among these simple souls, and his fortune is pretty soon made; he plays with them.* Social welfare was part of the Christian ethos. It attracted Peregrine to the community. But again, it was something Lucian got wrong. Apollos was an existentialist. The Christians were existentialists. They could be lied to and be made fools of. They gave Peregrine the benefit of the doubt. When they realized he was taking advantage of them their attitude toward him hardened. When he became captive audience, they bribed the jailers for a turn at preaching to him.

Lucian admired Apollos who gave him reason for admiring Christ. The God of War did not come into it. Reincarnation did not come into it. Christ was *distinguished personage.* Christ was charismatic because he was not the servant of the Lord. There was no question of him being anybody's servant. If he was a servant to anything it was to righteousness. Greek Christianity was different to Jewish Christianity. Paul the wannabe apostle admitted this. He said, *the Greeks seek after wisdom: But we preach Christ crucified, unto the Greeks foolishness.* No one can accuse Paul of wisdom. When Lucian said, *The Christians, you know, worship a man to this day* it was the same as saying they worshipped the literal Christ, and were all about the literal Christ.

Apollos worshipped Christ like he was deity and Christ acted like he was deity. Christ believed he was Yahweh the God of War, and he did a very convincing job of it. Christ was to all intents and purposes the God of War. In Christ the Lord was once again the lawgiver. When Lucian said *all this they take without evidence* he was not referring to the Holy Trinity or the Holy Doctrine of Divine Omnipresence or the Holy Spirit or the Logos or even the primary substance. Classical theism, creationism, and spirituality did not come into it. Apostles saw miracles of healing but that does not prove someone is the God of Ancient Israel, the same who drowned Egyptians and their horses in the Red Sea.

Miracles of healing were sensational but not evidence that Jesus was over one thousand years old and an ethnic cleanser.

Miracles came and went with Christ. Pliny's and Lucian's Christians were not about miracles or demons or praying or faith. The political association and social revolution did not stray into the supernatural. Their assessment was that God is a natural phenomenon. Never mind evidence the worship of a man defied logic. What they believed was beyond evidence. The simple souls and misguided creatures believed whatever they were told by their leaders or so it seemed. Apollos hated theologians and avoided them like the plague. Theologians could be relied on to spiritualize Christ in the same way they spiritualized Yahweh.

Lucian, Peregrine, Pliny, and Apollos provide the Greek perception of Christ. It is the "wisdom" perception. Pliny wrote an exhaustive account of what they were about. It does not include the supernatural because the supernatural did not exist for them. The New Testament books of Romans and Corinthians are very critical of the Greek super apostles. They express great loathing for Apollos a real-life friend of God. His was not delusion or an invisible friend. He was conscious of God as God is. His consciousness betrayed the fact that God is a man, his religion did the same. His head was clear of Hellenistic Judaism. Peregrine's leap into fire sparked Lucian's interest in the Christians. He found a beautiful Greek religion.

Peregrine was drawn to the famed generosity of the Christians. Lucian thought they invited exploitation. *Now an adroit, unscrupulous fellow, who has seen the world, has only to get among these simple souls, and his fortune is pretty soon made; he plays with them.* Peregrine did well out of the Christian communities of Turkey. But it was not all one way. They did well out of him. He used to be a Cynic preacher. He was an academic, an educator, and a spiritual guide. He started a trend because other Cynics became less suspicious of the Christians. It was something they could try out. Peregrine loved being the center of attraction. He was a showman, and he was Christian only for show. But that was precisely what the Church wanted of him, he put Christianity on show.

The infamous Socratic philosopher and theologian was an advertisement for the Christian movement. They needed him. They wanted to become mainstream. Converts needed to take their conversion a lot more seriously and commit to the movement long term. Christians lacked depth and knew little of divinity. Peregrine offered a lot of depth. Peregrine had people talking about the religion no one could understand. The church wanted to grow, and Peregrine had a lot to offer. The fact that he was a bit of a crook made him a challenge for salvation. According to Pliny one of the novel rites was to take an oath not to commit fraud or refuse to return a trust when called upon to do so. Assimilation was on the agenda for both parties.

Peregrine was not taking a backward step. He was not going to worship a man. It was a religion where unlearned men and women were taking up leadership roles. They did not pray nor make sacrifice to any of the gods. They were about celebrating Christ through hymns sung to him in thanks for their taste of heaven. As individuals and as a group they expected normal relations with Jesus Christ the man upon death and transfer to his off-world kingdom. There were a lot of gaps in their knowledge and logic. The Lord's Supper was the entry point into the consciousness of Christ, it put their Christian experience into context. They also considered it to be an entry point for Christ himself into their lives. The gathering was always conducted in his name and nobody else's.

Peregrine was a theologian and a reliable intellectual who did not want to mislead anyone into believing God is a man. He was a Cynic. He was also there for personal gain. The gifted individual thought the world owed him a living. The "unscrupulous" fellow played with the simple souls and made his fortune. The misguided creatures brought exploitation upon themselves. He was the forerunner of today's tele-evangelists.

Sin and faith have always been great money spinners. He took their money. In return he brought them up to speed on theology. *He pretty soon convinced them of his superiority; prophet, elder, ruler of the Synagogue--he was everything at once; expounded their books, commented on them, wrote books himself. They took him for a God, accepted his laws, and declared him their president.*

The Christians acquired spiritual concepts. Peregrine managed to modernize the belief system. It would have been easy to explain invincibility as resurrection and dualism. The Lord's Supper was not a mechanism for dealing with sin. A crack was opened for Hellenistic Judaism to seep through. Christ crucified is a mechanism for dealing with sin. Faith was back to the top of the agenda and was needed seeing that no one understood how a man could also be God. Peregrine offered logic and not evidence. The evidence they had was circumstantial and anecdotal. It was not enough. Peregrine was an expert in the natural sciences and Intelligent Design. Greek Christianity was set in direction of the primary substance.

Peregrine enjoyed his time as a Christian. He was more famous than ever, and he went on tour with the story of his salvation. But he was a man who liked spending money and he enjoyed the high life. He could not stay out of debt. He was back in trouble with the law and found himself in prison. Captivity took the theologian from insincerity to hypocrisy. Lucian described what punishment did for the sinner. *This was the very thing to lend an air to his favourite arts of clap-trap and wonder-working; he was now a made man.* He played victim because it generated revenue. *The activity of these people, in dealing with any matter that affects their community, is something extraordinary; they spare no trouble, no expense.* His brothers had not given up on him. They appreciated his work in making them more credible in terms of philosophy and formal education.

Philosophy was the hobby of the governor of Syria, and he knew all about Socrates and his imprisonment and death and fame. The criminal was still being remembered six hundred years later. The modern-day Socrates was getting arrested all too often. His brushes with the law made him famous. The law was become Peregrine's publicity machine. It got to the point that the governor refused to punish him. *He discharged him--wouldn't hear of his being punished--and Peregrine returned to Armenia.* The governor was one person who was not entirely surprised by the manner of his death. Peregrine succeeded because he is remembered for his death, a death which happened to reveal the state of Greek Christianity.

It is because of Peregrine that finally the intelligent classes were curious about the Christian movement. They used to dismiss it as silliness because the flesh does not survive death. The Lord's Supper was a remembrance service for someone who was not dead. They worshipped a man because they did not have a God. It was the world of the bizarre. Lucian lost a friend, and he was angry. Peregrine provided the traction into the world of the Christians. Their effect on him and his effect on them made for the story. Research for the story brought Apollos to the attention of Lucian. He found a Greek superhero buried in the past. There was also a fabulous, distinguished *personage* around whom the Greek built a church.

Christians betrayed their past for a theological appreciation of themselves. Lucian had a theological appreciation because of his fondness for the works of Plato. He had no desire to betray his past for a Christian perception of life and the world. He would need to achieve decreased consciousness through intellectual discipline. Christians did not have his education and the religion seemed to be designed for people without formal education. Spiritualization of Christ had not taken place when Lucian declared, *The Christians, you know, worship a man to this day.* But to get things into perspective, an improved social environment led to better mental health. The lifestyle increased life expectancy.

The Christians had a beautiful thing in their revolving door. People were free to try out the religion and leave if not satisfied. Lucian could have and would have attended a meeting or two. In his own way he was familiar with those who abandoned Peregrine to his fate. He was out to discredit them. He said *they despise all worldly goods alike, regarding them merely as common property.* Impractical and possibly unethical by Lucian's standards. Common property was a manifestation of brotherhood. It was the kind of spirituality the unscrupulous Peregrine liked. He did well out of it which was why he made them despise worldly goods even more. The original narrative was preserved in the hymns they sang. The original narrative was safe. The Cynic was able to introduce subtle changes.

Money was not required for animal sacrifice. There was not a temple or a priesthood to be maintained. All a person needed for entering the gathering was curiosity. Peregrine the Cynic tried it out, but it was not for him which was fair enough. No one in history has been more divisive than Yahweh. There was nothing universal about the God of War. Theirs was a personality cult, and they had a social consciousness which did not appeal to everyone. Leaving the church was all too easy. The problem was the lies coming from the likes of Lucian and from the temple cults and schools of philosophy. Existentialists do not despise all worldly goods alike. Christians were about sharing existence in the same way that God was about sharing existence.

Peregrine was not someone who liked sharing. He was a Cynic before and after his incursion into the simple social association of the Christians. He died a Cynic. He leapt into fire on the understanding that as an act of faith it qualified him for afterlife. Faith was his get out of jail card. One brilliant act of faith was going to make him righteous and justify his whole existence. Otherwise, he was a con artist, a fraud, a liar, and a cheat. He expected to be rewarded for his faith. He etched himself on the memories of all those who witnessed him burn. They watched him pass into the afterlife. Peregrine did not just want to be remembered forever. He wanted it remembered forever that he was a Cynic.

Peregrine's death was to be a testimony of his belief in spirituality. He had no idea that Jesus Christ died for his sins because that was not something which the Greek Christians believed or encouraged. Christianity could not save the Cynic from sin. There was no mention of sin in Pliny's court report. Pliny understood the simple souls. They were moral beings. They took an oath *not to commit fraud, theft, or adultery, or falsify their trust.* That prepared them for the day's temptations because oaths translated into behaviour. Pliny spoke their language and offered them pardon through repentance to avoid the wrath of Rome. Peregrine had to choose between a leap into fire or a lifelong struggle to be a person of good character.

Instead, he died a death to prompt spiritual thoughts in those attending the Olympics. It was to remind them that there was more to life than physical accomplishments. They were reminded to be mindful of the spirit. Peregrine was a theologian through and through. He could not help but bring theology and spirituality to simple souls. Apollos turned his back on theology. He was an existentialist. He got to observe God closely. He observed existentialism in action. It rubbed off on him. Apollos was that much more conscious of the switch from theology to existentialism because of his education. He could distinguish the theological from the non-theological.

Primary substance practitioners believed "*All things arise from a primary substance and return to it, unless they pay fine and penalty for their iniquity*". Militant atheism discovered Yahweh the God of War. A militant atheist discovered

Yahweh-Christ. An existentialist is someone who lives in the same world as philosophers but sees things differently. A primary substance practitioner has intellectual discipline and will not say Yahweh for fear of blasphemy and eternal damnation. An atheist will not say Yahweh-Christ if there is risk of lending support to either a creationist or a skygod worshipper.

An existentialist will say Yahweh because Pontius Pilate made it lawful to call someone by that name. It was Yahweh-Christ the God of War who entered the Garden of Gethsemane and later fell to the ground. It was Yahweh-Christ who used to be the God of War who left the garden to face justice and due process. He got his name back courtesy of the Roman magistrate and he was happy. But the reformation was about Christ and choosing between the historical and the philosophical. Shakespeare's plays favoured Christ of history. He said, "There are more things in Heaven and Earth, Horatio, than are dreamt of in our philosophy."

www.ingramcontent.com/pod-product-compliance
Lightning Source LLC
LaVergne TN
LVHW050817061126
851857LV00012B/843
* 9 7 8 1 9 1 6 2 7 0 0 6 0 *